Robotic Systems: Modelling, Technology and Applications

Robotic Systems: Modelling, Technology and Applications

Edited by **Rowland Wilson**

WILLFORD PRESS

New York

Published by Willford Press,
118-35 Queens Blvd., Suite 400,
Forest Hills, NY 11375, USA
www.willfordpress.com

Robotic Systems: Modelling, Technology and Applications
Edited by Rowland Wilson

International Standard Book Number: 978-1-68285-087-9 (Hardback)

Contents

Preface

Robotics is a vast and dynamic discipline which has witnessed ample progress in the past decade. This book explores all the important aspects of robotic systems in the present day scenario. Significant topics in this field such as mechatronics, control and modeling of robotic systems, human-machine interaction, artificial intelligence in robotics, etc. have been extensively discussed in this text. It strives to provide a fair idea about this discipline and to help develop a better understanding of the latest advances within this field. This book includes contributions of experts and scientists which will provide innovative insights into this field. For all students who are interested in robotics, the case studies included in this book will serve as an excellent guide to develop a comprehensive understanding of the subject.

This book is a result of research of several months to collate the most relevant data in the field.

When I was approached with the idea of this book and the proposal to edit it, I was overwhelmed. It gave me an opportunity to reach out to all those who share a common interest with me in this field. I had 3 main parameters for editing this text:

1. Accuracy – The data and information provided in this book should be up-to-date and valuable to the readers.
2. Structure – The data must be presented in a structured format for easy understanding and better grasping of the readers.
3. Universal Approach – This book not only targets students but also experts and innovators in the field, thus my aim was to present topics which are of use to all.

Thus, it took me a couple of months to finish the editing of this book.

I would like to make a special mention of my publisher who considered me worthy of this opportunity and also supported me throughout the editing process. I would also like to thank the editing team at the back-end who extended their help whenever required.

Editor

A Test Platform for Planned Field Operations Using LEGO Mindstorms NXT

Gareth Edwards *, Martin P. Christiansen, Dionysis D. Bochtis and Claus G. Sørensen

Department of Engineering, University of Aarhus, Blichers Allé 20, Tjele 8830, Denmark;
E-Mails: martinp.christiansen@agrsci.dk (M.P.C.); dionysis.bochtis@agrsci.dk (D.D.B.);
claus.soerensen@agrsci.dk (C.G.S.)

* Author to whom correspondence should be addressed; E-Mail: gareth.edwards@agrsci.dk

Abstract: Testing agricultural operations and management practices associated with different machinery, systems and planning approaches can be both costly and time-consuming. Computer simulations of such systems are used for development and testing; however, to gain the experience of real-world performance, an intermediate step between simulation and full-scale testing should be included. In this paper, a potential common framework using the LEGO Mindstorms NXT micro-tractor platform is described in terms of its hardware and software components. The performance of the platform is demonstrated and tested in terms of its capability of supporting decision making on infield operation planning. The proposed system represents the basic measures for developing a complete test platform for field operations, where route plans, mission plans, multiple-machinery cooperation strategies and machinery coordination can be executed and tested in the laboratory.

Keywords: field robots; indoor simulation; micro-tractor; operations management; area coverage

1. Introduction

Full-scale testing of agricultural operations management can often prove both costly and time consuming, while computer simulations often make assumptions and estimates about the environment, sensors and actuators in the system. In particular, when considering agricultural operations, full-scale testing can only be carried out at certain times of the year, possible only a few months, and tests on the same area cannot be easily repeated, *i.e.*, a crop can only be harvested once.

Computer models intended to simulate sensors and actuators are only a representation of reality with a certain level of accuracy. The models are designed to simulate scenarios the developers have deemed relevant to test design parameters. In [1], GPS signals are simulated to realize the external noise sources affecting the operations of an agricultural vehicle's auto-steering system. The GNSS and vehicle model are tested with a nonlinear model predictive controller. The current system models are still only designed to test the scenarios the developers want to research based on current domain knowledge.

Software tools for modeling and simulation of robot vehicles exist in the form of tools, such as player-stage-gazebo and Microsoft robotics studio. Game engines for physical simulation or model-based differential equations allow a robotics simulation tool to simulate the system physics [2]. Robotics simulation frameworks have been used to move directly from simulation to full deployment on a vehicle. Robotics simulation frameworks provide a number of generalized building blocks (vehicle, sensors and actuators) that can be modified to describe different setups. To select viable solutions, extensive domain knowledge of the system type and tool building blocks is needed. The authors of [3] first use computer simulation and then real life testing to gather results on the effectiveness of a system to control small robots during an environment discovery procedure. Simply procedural algorithms were tested in the computer simulation, and once their robustness was proven, real life testing was carried out on a small scale.

An intermediate step between simulation and deployment has been developed in recent years, by utilizing a Hardware-In-the-Loop [4] test setup to evaluate an algorithm's control response and robustness. A Hardware-In-the-Loop test setup is still dependent on the correct modeling of sensors and actuators, for evaluation of the control loop.

In the case of field machinery operations, whilst there are a number of examples for the implementation of test platforms and small-scale machines, these are limited. The authors of [5] used two iRobot Magellan Pro robots in an indoor environment in order to demonstrate a methodology for real-time docking of combined harvesters and transport carts. The authors of [6] used the iRobot platform to test a swarm intelligence algorithmic approach for multi-robot setup for controlling weed patches distributed within a field area, and [7] developed a robotic platform equipped with cameras for row guidance and weed detection for the mapping of weed populations in fields, which was used to demonstrate intelligent concepts for autonomous vehicles.

Nevertheless, the above-mentioned examples are customized tools developed specifically for each application under study and do not build in a common standard framework.

LEGO Mindstorms is an example of a common framework that has been used in other scientific disciplines related to robotics, e.g., robotic exploitation [8] and team intelligence [9]. LEGO Mindstorms provides a proven, versatile framework for prototyping mechanical robotic systems that are programmed with a high degree of complexity. It also provides a system that has the ability to add and remove functionalities, as well as to reconfigure its architecture. This allows it to adapt to the needs of the different requirements of various applications, giving it an advantage over other frameworks. This critical notion is in accordance with the requirements of future innovative agricultural fleet management systems, as have been outlined [10].

In order to quickly test operational management techniques, a test platform was developed, utilizing a LEGO Mindstorms micro-tractor, allowing for easily replicable results that can be evaluated while interpreting collected data. The test platform also consists of control and display modules that enable it to execute and monitor management techniques. Compared to a Hardware-In-the-Loop, solution the micro-tractor allows for the evaluation of software components using actual sensory input. This test platform is seen as an intermediate between simulation and full-scale testing, rather than a replacement of either.

In this paper, the test platform is described in terms of its hardware and software components. The performance of the platform is demonstrated and tested in terms of its capability of supporting decision making on field operation planning by indoor environment simulations. Following this introduction, the LEGO Mindstorms suite is described in Section 2. In Section 3, the hardware and the software components are described. Section 4 outlines the tests, which were conducted to prove the test platforms' fitness for the purpose, and finally, conclusions are made in Section 5.

2. The LEGO Mindstorms NXT

LEGO Mindstorms is a suite developed by LEGO containing the "NXT Intelligent Brick" as the main controlling unit. It is programmed either using LEGO's own Mindstorms IDE (integrated development environment) or various third-party development tools. The NXT Brick is capable of controlling three LEGO NXT servo motors in terms of rotation speed and direction, via voltage regulation. The NXT servo motors also have built-in rotary encoders that can deliver 720 steps, equivalent to an accuracy of 0.5°, which are used to monitor the angular position respective to their starting position, which is deemed to be zero degrees. The NXT Brick can have up to four sensors as inputs through either analogue or I2C connections. These sensors include standard LEGO sensors, such as light sensors, touch sensors and ultra-sonic sensors, and sensors developed from other companies (e.g., ViTech, Microinfinity, Dexter Industries), such as temperature sensors, color sensors, chemical sensors, *etc.*, coping with the measuring requirements of scientific experimentations.

3. Methods

3.1. Hardware

The steering of the tractor is actuated with a rack and pinion system, which allows the front wheels to turn through ±30°. A standard NXT motor was used to control the steering (Figure 1a) with a gearing at a ratio of 7:1 to increase the range of the control. The rear wheels are controlled by another NXT motor (Figure 1b), which transmits the power to the back axles via a differential gear. This allows the vehicle to turn corners without the back wheels slipping. The specific relation of the gearing ratio and the size of the rear wheels tires results in a 0.51 mm movement of the tractor for each degree that the drive motor turns.

The micro-tractor was designed to be a representation of a generic tractor, rather than a specific tractor, so as to allow more flexibility in the transferability of the results. The micro-tractor has a wheelbase of 175 mm and a turning radius of 370 mm. Considering that an average medium-sized tractor (150 hp) has a wheelbase and turn radius of approximately 2.5 m and 5.2 m, this would correspond to a scaling of 1:14. If there is a need for the test result to demonstrate a specific tractor, the use of LEGO would allow for fast modification.

Figure 1. Photos of the steering and drive components.

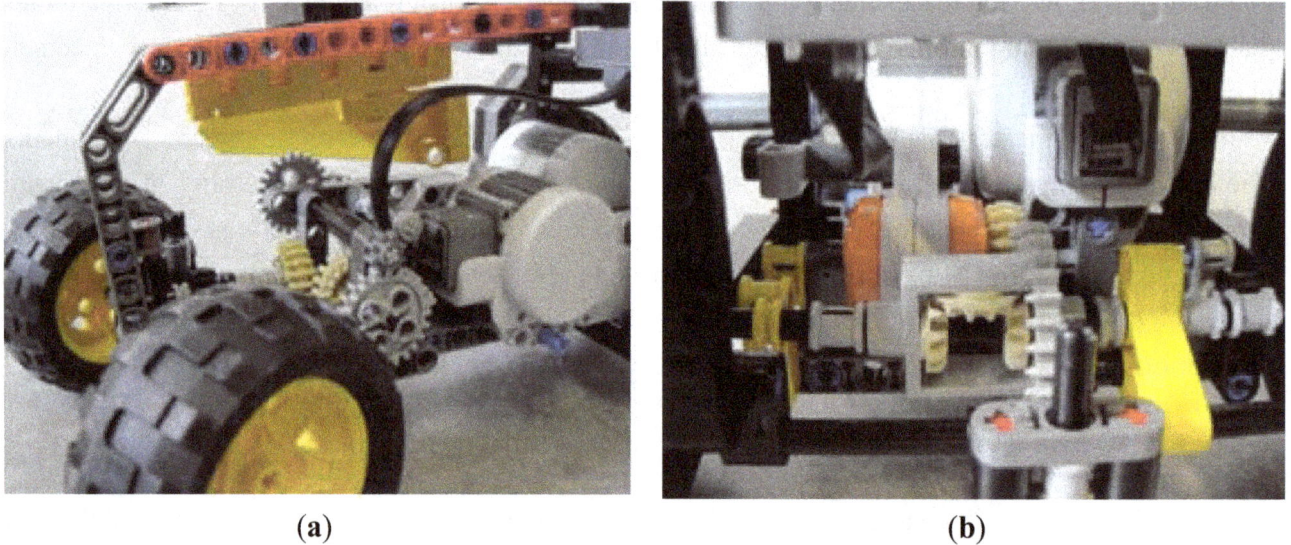

(a)   (b)

The main navigation sensor is the CruizCore® XG1300L IMU, which is mounted on the front of the micro-tractor and is able to measure the relative heading of the micro-tractor compared to the starting position with a relative accuracy stated as <0.1°. The device contains a single axis MEMS gyroscope and a three-axis accelerometer. The signals from these sensors are processed onboard the device, applying factory set compensation factors, which helps to reduce the most significant errors. The measured heading is susceptible to a maximum error of 10°, according to the product specifications, during one hour of continuous operation.

As part of developing a platform to demonstrate various agricultural operations, implements can be constructed using additional NXT units. However, the micro-tractor has one motor port and three sensor ports available for implements that are not equipped with an NXT unit. The micro-tractor is equipped with a drawbar suitable for connecting implements.

3.2. Software

The BricxCC (Bricx Command Center), an open source Windows program that uses the NXC programming language [11], is used to compile the programs contained on the NXT Brick. Matlab (MathWorks®) and the RWTH-Mindstorms NXT toolbox [12] were used for remote communication with the NXT Brick via Bluetooth (Figure 2).

Figure 2. The communication architecture.

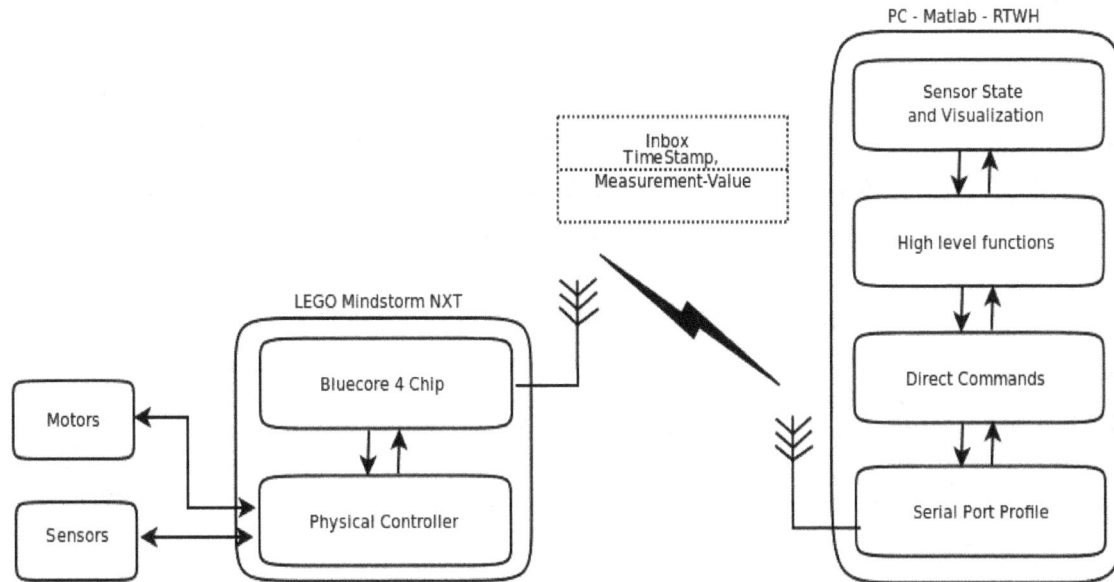

3.2.1. Communication

The Bluetooth protocol utilized by LEGO is placed on top of the Serial Port Profile (SPP) protocol. Direct control commands provide the ability to remote control the NXT from a computer. Each NXT command over a Bluetooth connection takes approximately 100 ms to successfully process, making it too slow for precision control of the tractor. As a consequence, the micro-tractor was chosen to be programmed in the NXC programming language, and the compiled code was loaded directly onto the NXT for execution. Programming the NXT directly provides the ability to control the position and angle with a much higher accuracy, compared to the Bluetooth solution. If communication between the computer and brick is lost at any point during the testing, the NXT makes a sound, so that testing can be aborted and restarted.

3.2.2. Route Planning

The route planning for the micro-tractor was implemented offline using the Matlab programming language. The input for planning includes the boundaries of the working area, which can be selected in a digital map, and a number of operational parameters (Figure 3). Based on the input, as the first step, the geometrical representation of the field is generated. The geometrical representation regards the definition, in terms of their coordinates, of the geometrical entities inherent in a field area representation. These entities include the parallel field-work tracks and the peripheral boundary passes (headland area). The next step includes a coverage path generation, which could be either a conventional plan (e.g., sequential ordering of the tracks) or optimized according to the principle of *B-patterns*, that algorithmically results in an optimal field-work track traversal sequence according to an optimization criterion [13,14]. In the latter case, the coverage plan does not follow the repetition of standard motifs, but the plan is a unique result of the optimization approach on the specific combination of the mobile unit kinematics, the operating width and the optimization criterion, such as, total or non-working travelled distance, total or non-productive operational time, a soil compaction measure [15], *etc*. In the presented case, the non-working travelled distance has been considered as the minimization criterion. The optimization problem is that of finding the optimal track sequence: $\sigma^* = \arg\min_{\sigma} \sum_{i=1}^{|T|-1} c_{p^{-1}(i+1),p^{-1}(i)}$, where $T = \{1,2,3,...\}$ is the arbitrarily ordered set of the field tracks that cover the entire field area, $\sigma = < p^{-1}(1), p^{-1}(2),..., p^{-1}(|T|) >$ is a permutation, $p(\cdot) : T \to T$ is the bijective function, which for any field track $i \in T$ returns the position of the ith field track in the track traversal sequence, and $c_{p^{-1}(i+1),p^{-1}(i)}$ is the cost for moving between tracks, $p^{-1}(i+1)$ and $p^{-1}(i)$, which, in the particular case, corresponds to the nonworking travelled distance.

Figure 3. The architecture of the route planning.

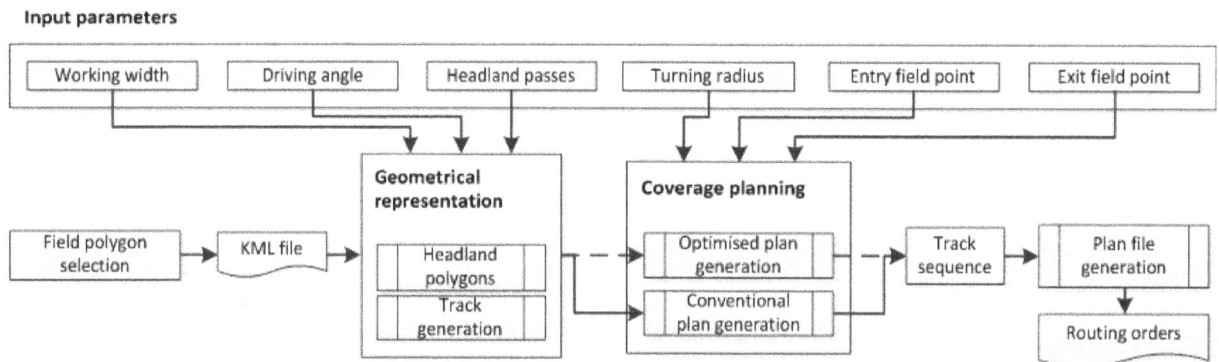

The final function is the generation of the routing orders, which include a sequence of straight lines and turnings executions. Straight line segments are described by the heading, the distance to be travelled, the driving speed and the starting X and Y coordinates. Turning segments are described by the initial heading, the final heading, the direction of the turn (clockwise or anti-clockwise) and the driving speed.

3.2.3. Position Determination

The micro-tractor determines its position onboard the NXT using the heading value from the IMU and the encoder value from the drive motor. Using these values, the position and heading are calculated relative to the micro-tractor's starting position and heading. While communicating with the visualization computer, the micro-tractor samples the IMU and drive motor encoder and calculates its position, at a rate of approximately 12 Hz. Using these techniques for position determination requires the micro-tractor to be operated on a face surface with minimal slip between the wheels and surface occurring.

3.2.4. Vehicle Navigation Control

The route maintains its structure of a straight line and turning segments. The segment commands are passed to the NXT one at a time from the Matlab control system; this allows for the execution of management techniques that require real-time adaption of the route. During straight segments, the NXT calculates the number of revolutions of the drive motor it needs to make to drive the prescribed distance. While this is executing, the NXT monitors the micro-tractors distance from the line normal to the direction of travel and the angular error in the heading to that direction of travel (Figure 4). These two calculated errors are entered into a transfer function, and the NXT makes an adjustment to the steering wheels in order for the micro-tractor to reduce these errors and follow the line as described. A similar control system is described in [16] for use with a full-scale four-wheeled machine, where the errors are referred to as the lateral and angular error. The LEGO test platform also assumes that it is operating on a hard, flat surface with minimal slip. The parameters for the transfer function used were determined empirically.

Figure 4. Heading error and distance from the line to travel.

To execute a turn segment, a second control function is used. The micro-tractor sets its wheel in a full lock position in the direction of the turn and then starts the drive motor. During the turn, the heading is monitored until the micro-tractor reaches its desired angle, at which point the drive motor is stopped and the steering wheels are turned back to the zero position. The reason the steering wheels are moved while the vehicle is stationary is to ensure that the micro-tractor traces perfect circles.

3.2.5. Visualization

The estimated current heading and position of the micro-tractor are written to a text string and passed into separate mailboxes with 100 ms division, overwriting the old message in the mailbox, along with a timestamp. The task of the Matlab system is to read the content of the mailbox and store and display the results. The current state of the tractor is then calculated and plotted on the map, the travelled path and the desired path are also plotted for comparison reasons.

3.3. Test Platform Architecture

The architecture of the LEGO test platform aims at mimicking the real-world system in a meaningful way (Figure 5). The main three modules of the system are the Position Determination, Vehicle Control and Visualization. Each of these modules is replicated within the test platform. Within the LEGO platform, although the Route Planning and Visualization are separate systems, they are run on the same computer. The dashed lines on Figure 4 indicate the components of the system that contain the modules. There are some differences between the component setups in the systems; however, this does affect the functionality. For example, the connection between the Vehicle Control and Visualization modules is implemented via a wired connection in the real-world system and a wireless, Bluetooth connection in the LEGO platform. The functionality of these connections is simply to pass information from the Vehicle Control module to be displayed by the Visualization module. The rate at which this information is sent, approximately 10 Hz, is well within the tolerance of the Bluetooth connection; plus, as mentioned in Section 3.2.1, if the connection is interrupted, the test is aborted. Therefore, the Bluetooth connection has the same functionality as the wired connection.

A similar full-scale testing system is described in [17]. Route plans are first generated on a computer and transferred to the test tractor via USB. The tractor then executes the plan, while performing vehicle navigation, and displays the results on a small onboard computer. By using a similar architecture on the test platform as in the real world, the solutions that are found, such as route plans and management techniques, are able to be transferred to the real-world system more effectively.

Figure 5. Depiction of LEGO platform and real-world architectures.

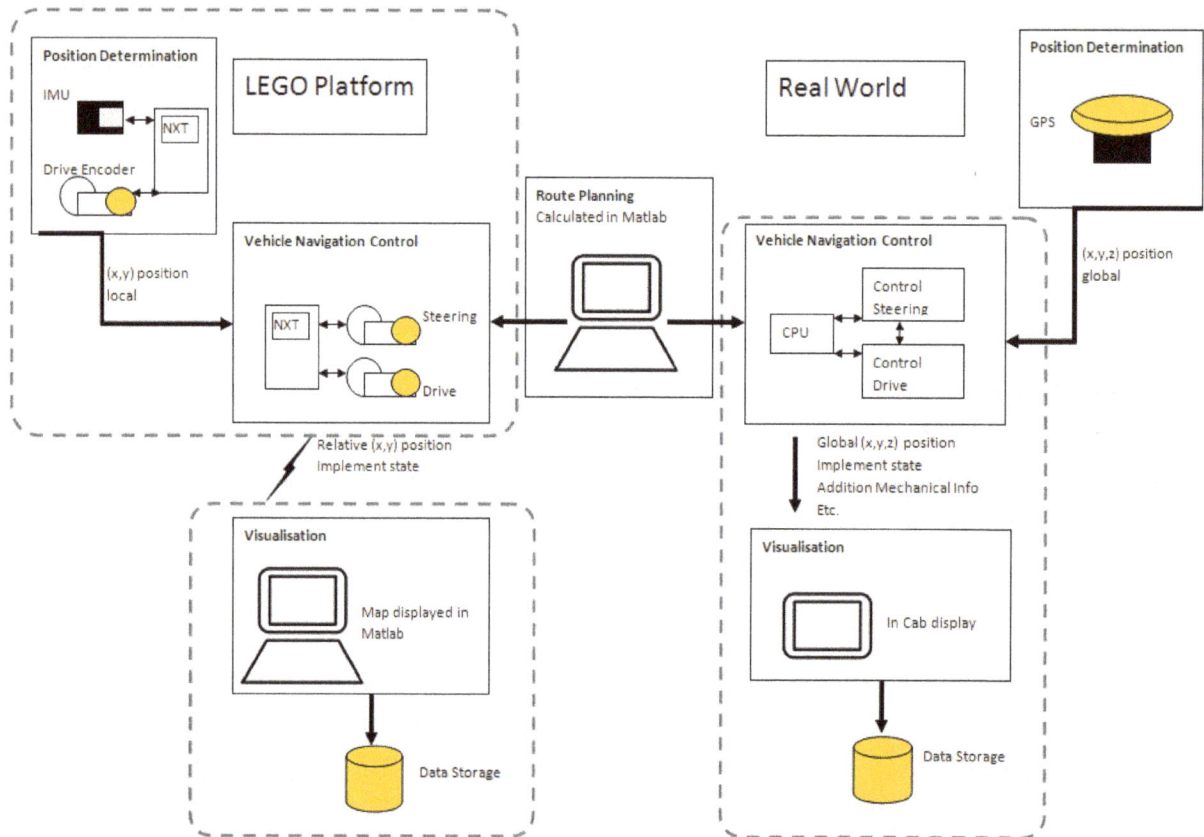

The methods used in the Position Determination modules of the LEGO platform and the real world are vastly different; however, their outputs are the same. A limitation of the current Position Determination module in the LEGO platform is that the operation surface must be flat and provide minimal tire slip, which is not the case in the real world. A limitation of the real-world GPS system is the need for contact with many satellites, which can be susceptible to overhead obstructions, such as trees or cloud cover. Since the LEGO platform operates indoors, the use of a GPS system would be extremely difficult. In the real world, a combination of sensors, such as computer vision techniques or multiple GPS antennae, would be required to obtain an accurate estimate of the vehicle's heading; however, in the LEGO platform, the IMU sensor is sufficient. In both systems, the Position Determination modules provide the Navigation Control module with an estimation of the current position and the current heading, so that steering corrections can be made, and in this way, they can be considered to be comparable.

The system architecture of the platform is built in a modular manner, so that components, such as the Navigation Control or Visualization, could be easily exchanged with another module, as long as the new module takes the same inputs and gives the same outputs. To increase the functionality of the system to allow for real-time operations management, the link between the Route Planning and Navigation Control modules should be modified to a two-way connection, so that data can flow between them. This connection would allow the Route Planning module to update the current plan due to any changes that may be observed. The micro-tractor can receive commands to execute each segment of the path separately; therefore, the remaining segments of the path, after the current segment, are still open to being altered. This modification of the architecture will be investigated in future work.

4. Implementation of the Test Platform

4.1. Position Accuracy

An indoor GPS (iGPS) was used to test the accuracy of the micro-tractor position determination. The iGPS system (Nikon Metrology, NV Europe) combines a transmitter sensor placed at the center of the rear axle of the micro-tractor (Figure 6a) and six beacon posts (Figure 6b) located around the working area. The author of [18] documented the iGPS system capabilities to track movement up to 3 m/s with an accuracy of 0.3 mm. Opposite planar and angular motions were tested to ensure an unbiased dataset for evaluation. This confirms that iGPS is usable for both static and kinematic spatial positioning and tracking. The kinematic measurement mode of the iGPS was used to track the movements of the micro-tractor with a frequency of 40 Hz.

Figure 6. (**a**) The micro-tractor with mounted iGPS sensor and power source trailer; (**b**) the iGPS beacon.

(**a**)   (**b**)

A series of navigation accuracy tests were performed in "virtual" fields for different combinations of operating width and driving directions. For example, Figure 7 presents the three paths (off-line planned, on-line estimated and actual measured) on a "virtual" field for the case of a 250-mm working width and 0° driving direction. Based on the tests, for a basis driving distance of 71.43 m (corresponding to 1 km full-scale distance), including straight line driving (operating on a field-work track) and 180° maneuvering (headland turnings), the average cross-track error between the estimated path and the iGPS path executed by the micro-tractor was 0.028 m, corresponding to 0.39 m full-scale cross-track error. This error is comparable with a typical error in field machinery navigation based on a standard GPS system, thereby showing that the Position Determination is satisfactory for use.

In order to simulate the capabilities of RTK- and DGPS-based navigation systems, the accuracy of the proposed system would need to be increased. However, the improved system should be low-cost and flexible, which excludes the use of precise, but expensive systems, such as iGPS; therefore, further examples were executed using only the tractor's position determination. The inclusion of a more accurate position determination, once developed or sourced, would be relatively simple, due to the modular setup of the architecture described in Figure 5.

Figure 7. The planned path (the black line), the estimated path by the micro-tractor internal sensors (red line) and the actual path recorded by the iGPS (blue line).

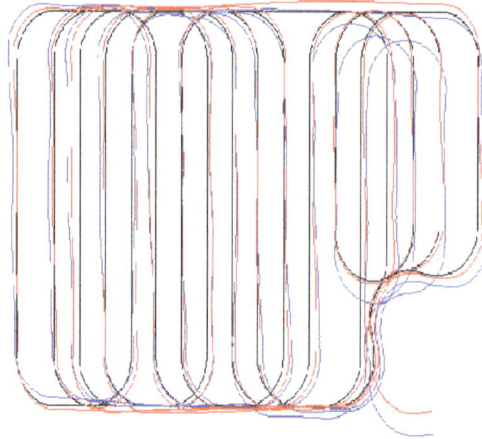

4.2. Demonstration Examples

To demonstrate the capabilities of executing and evaluating routing plans, the test platform was used to test area coverage plans with different setup parameters, such as working widths, driving angles, number of headland passes, *etc*. Figure 8 presents the executed plans from different operational scenarios on the same field. The accuracy of the micro-tractor's ability to maintain the predefined paths are detailed in Table 1, each test scenario was executed three times by the micro-tractor, and the results were then averaged.

Figure 8. Four different scenarios show prescribed route (black line) and driven route (red line) for working width and driving angle (**a**) 0.650 m—90°, (**b**) 0.8 m—90°, (**c**) 0.5 m—0° and (**d**) 0.25 m—120°. The axes are in the micro-tractors-scale and are in mm.

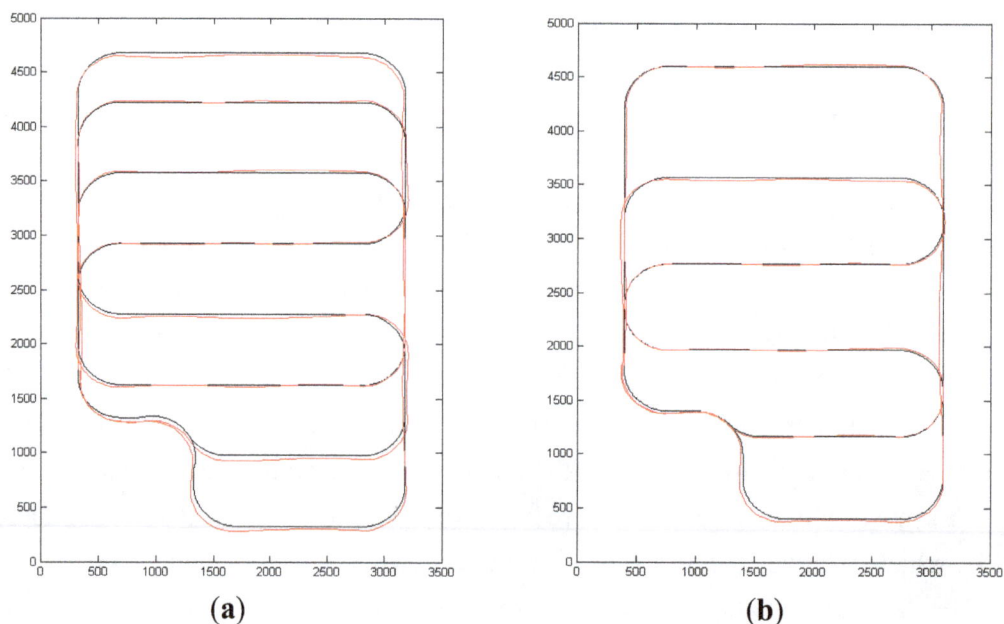

(a)   (b)

Figure 8. *Cont.*

(c)

(d)

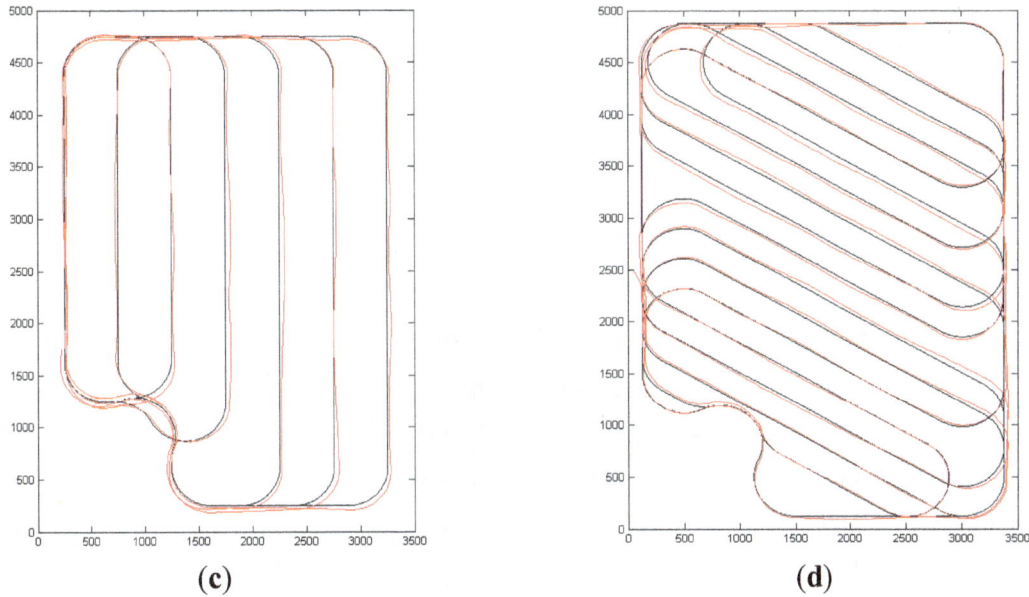

Table 1. Results from test scenarios.

Working Width (mm)	Driving Angle	Expected Path Length (m)	Actual Path Length (m)	Path Length Error	Average CPD * (mm)	Shown in
250	120	82.95	83.10	0.18%	26.76	Figure 8d
250	90	73.76	74.68	1.24%	24.15	
500	0	56.57	57.98	2.49%	20.25	Figure 8c
650	90	38.84	39.16	0.81%	20.27	Figure 8a
800	90	26.98	27.37	1.44%	15.92	Figure 8b
760	0	115.35	116.40	0.91%	52.03	Figure 9

* Cross Path Divergence.

To demonstrate the system's ability to test real-world scenarios, a real field was used in the final test scenario (Figure 9).

Figure 9. (**a**) The demonstration field; (**b**) predefined route (black line) and driven route (red line).

(a)

(b)

The field used is located at 52.42° N, 2.58° W, and the dimensions were sampled from Google Earth and scaled down by 1:14 to be relative to the test platform. The route was planned using a working width of 0.714 m (approximately, in full scale, 10 m) and driving parallel to the longest edge of the field.

5. Conclusions

The proposed system represents the basic measures for developing a complete test platform for field operations, where route plans, mission plans, multiple-machinery cooperation strategies and machinery coordination can be simulated and tested in the laboratory. The laboratory tests are easy to demonstrate and replicate at any time of the year; full-scale testing is often limited by weather and field conditions. Furthermore, using a small-scale test platform eliminated a lot of the safety concerns associated with operating large driverless machinery. The test platform should be seen as an extension to simulations-based evaluation rather than a replacement. Even the most stringently programmed simulations are susceptible to errors or to things being overlooked. The test platform provides another stage of quality assurance, with systems interacting with real collected data, before full-scale testing is attempted. The results from the test platform also add credence when planning full-scale testing and eliminate costly, superfluous tests.

The proposed system also provides many opportunities as an educational tool. Students can quickly and easily test management techniques in a classroom environment and see their ideas implemented in a physical way. Moreover, the modular setup of the system architecture allows students to develop and test new modules, gaining many insights into system engineering and controller design. While the use of the IMU and encoder values adequately estimates the micro-tractors position, it limits the system, as the surface it is run on must be smooth, flat and solid. For the purposes of testing operational management techniques, this is of no consequence, as fields are often simplified using flat 2D representation. However, if another method for efficiently determining the micro-tractor position were deemed necessary, this module could be replaced without affecting the rest of the system.

The execution of coverage plans was chosen to show the capabilities of the test platform to implement agricultural operations management techniques. The demonstration examples also show that is possible to evaluate coverage plan scenarios, involving different operational features (e.g., working widths, driving angles and number of headland passes), in terms of different operational efficiency measures, e.g., the measured non-working travelled distance, overlapped or missed area and operational time. The example using the real field (Figure 9) could be used as the first step in developing a Control Traffic System for the real-world field. Since the route planning has been shown to be effective, the full-scale testing could proceed more quickly.

The next steps are envisioned to be, for example, the inclusion of additional sensors (e.g., for mapping spatial variations within an area, crop row following, collision detection and enhanced navigation accuracy) and the implementation of multiple micro-tractor systems.

Acknowledgments

The authors would like to thank the Danish Ministry of Food, Agriculture and Fisheries for their funding of this research. The authors also want to thank Ole Green for initial supervision and support of the work. Thanks are due to Jens Kristian Kristensen at Research Centre Foulum for providing assistance in connection with the experiments. We would also like to acknowledge our colleague, Michael Nørremark, who provided input and expertise that assisted the research. We acknowledge Aachen University's Institute of Imaging and Computer Vision for providing the RWTH Toolbox.

Conflicts of Interest

The authors declare no conflict of interest.

References

1. Backman, J.; Kaivosoja, J.; Oksanen, T.; Visala, A. *Simulation Environment for Testing Guidance Algorithms with Realistic GPS Noise Model*; International Federation of Automatic Control: Winterthur, Switzerland, 2010; pp. 139–144.

2. Harris, A.; Conrad, J.M. Survey of popular robotics simulators, frameworks, and toolkits. *Southeast. Proc. IEEE* **2011**, doi:10.1109/SECON.2011.5752942.

3. Cepeda, J.S.; Chaimowicz, L.; Soto, R.; Gordillo, J.L.; Alanís-Reyes, E.A.; Carrillo-Arce, L.C. A behavior-based strategy for single and multi-robot autonomous exploration. *Sensors* **2012**, *12*, 12772–12797.

4. Rossmann, J.; Schluse, M.; Sondermann, B.; Emde, M.; Rast, M.; Advanced Mobile Robot Engineering with Virtual Testbeds. In Proceedings of the 7th German Conference, Munich, Germany, 21–22 May 2012; pp. 1–6.

5. Hao, Y.; Laxton, B.; Benson, E.R.; Agrawal, S.K. Differential flatness-based formation following of a simulated autonomous small grain harvesting system. *Citeseer* **2004**, *47*, 933–941.

6. Kumar, E.V. *A Swarm Intelligence Algorithm for Multi-Robot Weed Control an Emotion Based Approach*; Anna University Chennai: Tamil Nadu, India, 2008.

7. Bak, T.; Jakobsen, H. Agricultural robotic platform with four wheel steering for weed detection. *Biosyst. Eng.* **2004**, *87*, 125–136.

8. Kovacs, T.; Pasztor, A.; Istenes, Z. A multi-robot exploration algorithm based on a static bluetooth communication chain. *Robot Auton. Syst.* **2011**, *59*, 530–542.

9. Simonin, O.; Grunder, O. A cooperative multi-robot architecture for moving a paralyzed robot. *Mechatronics* **2009**, *19*, 463–470.

10. Sorensen, C.G.; Bochtis, D.D. Conceptual model of fleet management in agriculture. *Biosyst. Eng.* **2010**, *105*, 41–50.

11. Bricx Command Center. Available online: http://bricxcc.sourceforge.net/ (accessed on 30 August 2013).

12. RWTH—Mindstorms NXT Toolbox for MATLAB. Available online: http://www.mindstorms.rwth-aachen.de/ (accessed on 30 August 2013).

13. Bochtis, D.D.; Sorensen, C.G. The vehicle routing problem in field logistics part I. *Biosyst. Eng.* **2009**, *104*, 447–457.

14. Bochtis, D.D.; Vougioukas, S.G.; Griepentrog, H.W. A mission planner for an autonomous tractor. *Trans. Asabe* **2009**, *52*, 1429–1440.

15. Bochtis, D.D.; Sorensen, C.G.; Green, O. A DSS for planning of soil-sensitive field operations. *Decis. Support Syst.* **2012**, *53*, 66–75.

16. Oksanen, T. Path Following Algorithm for Four Wheel Independent Steered Tractor. Available online: http://cigr.ageng2012.org/comunicaciones-online/htdocs/principal.php?seccion=posters& idcomunicacion=12878&tipo=3 (accessed on 5 November 2013)

17. Blackmore, B.S.; Griepentrog, H.W.; Nielsen, H.; Nørremark, M.; Resting-Jeppersen, J. Development of a Deterministic Autonomous Tractor. In Proceedings of *CIGR* International Conference, Beijing, China, Novermber 2004.

18. Depenthal, C. Path Tracking with IGPS. In Proceedings of Indoor Positioning and Indoor Navigation (IPIN), Zurich, Switzerland, 15–17 September 2010; pp. 1–6.

2

The Role of Indocyanine Green for Robotic Partial Nephrectomy: Early Results, Limitations and Future Directions

Zachary Klaassen, Qiang Li, Rabii Madi and Martha K. Terris *

Department of Surgery, Section of Urology, Medical College of Georgia, Georgia Regents University, Augusta, GA 30912, USA; E-Mails: zklaassen@gru.edu (Z.K.); qli@gru.edu (Q.L.); rmadi@gru.edu (R.M.)

* Author to whom correspondence should be addressed; E-Mail: mterris@gru.edu

Abstract: The surgical management of small renal masses has continued to evolve, particularly with the advent of the robotic partial nephrectomy (RPN). Recent studies at high volume institutions utilizing near infrared imaging with indocyanine green (ICG) fluorescent dye to delineate renal tumor anatomy has generated interest among robotic surgeons for improving warm ischemia times and positive margin rate for RPN. To date, early studies suggest positive margin rate using ICG is comparable to traditional RPN, however this technology improves visualization of the renal vasculature allowing selective clamping or zero ischemia. The precise combination of fluorescent compound, dose, and optimal tumor anatomy for ICG RPN has yet to be elucidated.

Keywords: robotic partial nephrectomy; indocyanine green; near infrared fluorescence imaging

1. Introduction

Nephron sparing surgery (NSS) has been advocated as the surgical treatment of choice for the management of small renal masses, secondarily improving renal function and long-term survival [1]. Increasingly, robotic surgery has been utilized across many centers in the United States as the preferred NSS modality given the dexterity afforded by the robot and the ability of the surgeon to resect the mass and cover the renal defect. Particularly when tactile sensation is lost during robotic partial nephrectomy (RPN), the risk of incomplete tumor resection and positive margins can compromise cancer specific outcomes [2]. Intraoperative renal ultrasonography has historically been utilized to identify complex or intraparenchymal lesions [3], however the margin of masses may still be difficult to identify and may be cumbersome to use through a robotic assistant port.

The latest robotic surgery technology has attempted to circumvent these visual drawbacks, most commonly using near infrared fluorescence imaging, (*i.e.*, intravenous injection of indocyanine green (ICG) ([©]Akorn, Incorporated, Lake Forest, IL, USA)) to delineate perfusion and uptake of normal renal parenchyma during RPN (Figure 1) [4]. ICG binds to plasma proteins, remaining in the vasculature and providing excellent delineation of the vascular system [5]. Renal cortical tumors have reduced expression of bilitranslocase, a carrier protein for ICG in normal parenchyma proximal tubule cells, thus leading to a reduction in near infrared fluorescence imaging in these tumors and delineation from normal parenchyma [6].

Figure 1. Toggling between (**A**) white and (**B**) near infrared fluorescence (NIRF) to evaluate renal mass margins and (**C**) maintenance of equal fluorescence of normal renal parenchyma covering tumor and toward normal kidney. (Used with permission, Elsevier: Krane, L.S.; Manny, T.B.; Hemal, A.K. *Urology* **2012**, *80*, 110–118).

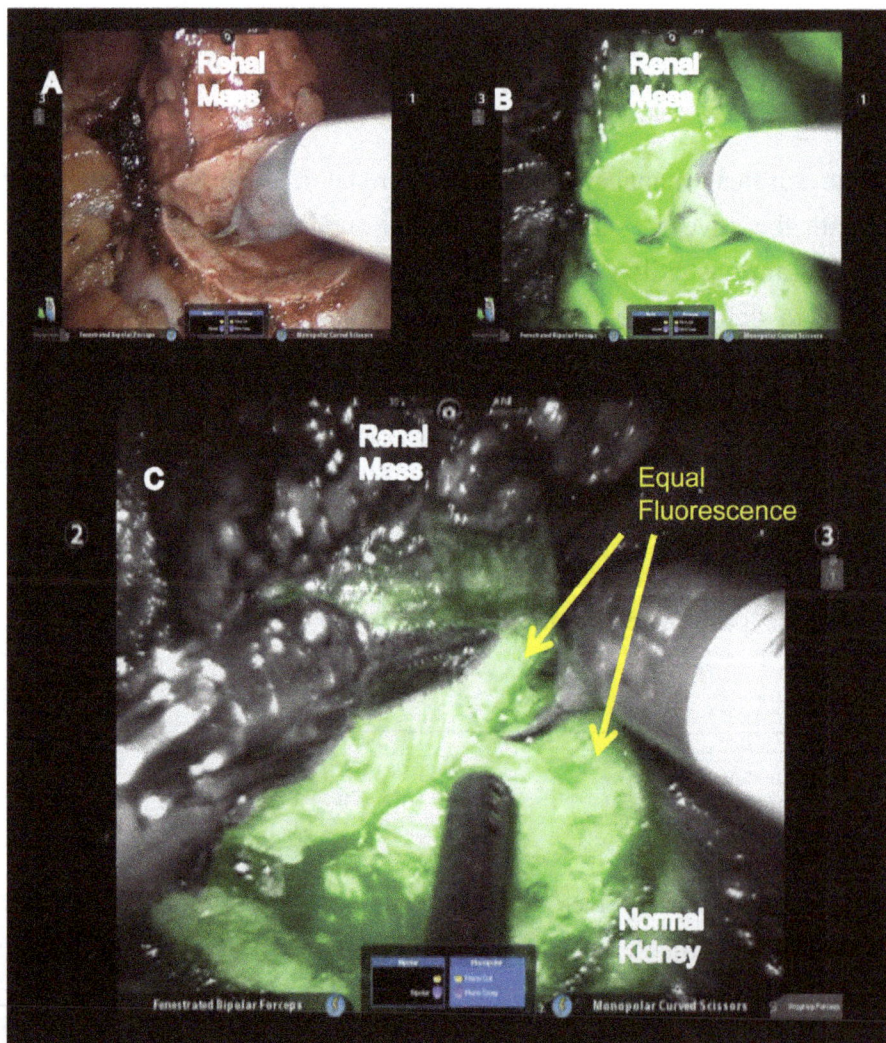

The purpose of this review is to analyze the current institutional experiences utilizing ICG during RPN, to address limitations and clinical situations when ICG may not be useful, and to highlight future directions of near infrared fluorescent imaging.

2. ICG RPN Institutional Studies

In 2011, Tobis *et al.* [4] were the first to publish their experience of ICG for RPN, having previously reported their success using ICG for open partial nephrectomy [7]. In their series of 11 patients, nine patients successfully underwent RPN and two patients were converted to robotic radical nephrectomy secondary to an intraoperative identification of renal vein thrombus and dense perinephric inflammatory adhesions. Median warm ischemia time (WIT) was 19.3 min, estimated blood loss (EBL) was 100 cc, radiographic tumor size was 3.8 cm, nephrometry score was 7.5, and robotic surgical console time was 134 min. Their technique for injection of ICG included preparing 2.5 mg/mL at the beginning of the case and injecting as a bolus dose within six hours of preparation. Tobis *et al.* used a total of 0.75 to 7.5 mg per injection based on the optimal tumor fluorescence, and injections were repeated as necessary to achieve adequate visualization taking care to remain below the recommended daily maximum dose of 2 mg/kg [4]. Fluorescence was seen in the renal vasculature between 5 and 60 s following injection in all patients and renal parenchyma was seen to fluoresce in 1 min (maximal fluorescence lasting 10–15 min) from injection. The tumors demonstrated decreased or no fluorescence compared to the surrounding parenchyma in 8 patients and equivalent fluorescence in 3 patients. On final pathology, 10 of the 11 masses were malignant and there were no positive margins; median change in post-operative compared to preoperative creatinine was 0.1 mg/dL.

Following these initial positive results, Krane *et al.* [8] reported their experience in 47 consecutive patients undergoing ICG for RPN, who were compared to 47 consecutive previous patients undergoing RPN without ICG. Both groups had comparative demographic (age and comorbidities), preoperative (kidney function) and renal tumor anatomy (nephrometry score and tumor size). Intraoperatively, the ICG group had decreased WIT (15 *vs.* 17 min, $p = 0.03$; only cases with hilar clamping included in calculating WIT) and more patients who underwent mass excision without hilar clamping (30% *vs.* 10%, $p = 0.0002$), however the ICG group had greater EBL (75 *vs.* 50 cc, $p = 0.023$) although likely not clinically significant. Length of stay, incidence of malignancy, estimated glomerular filtration rate (eGFR) and positive margin rate were comparable between the two groups.

Recognizing that under dosing ICG causes inadequate fluorescence of peritumor parenchyma and overdosing causes inappropriate tumor fluorescence, Angell *et al.* [9] reported their experience of 79 patients utilizing a dosing strategy to determine the optimal ICG dose. Their dosing strategy consisted of giving a minimum of two ICG doses including a test dose prior to complete kidney exposure and a calibrated second dose prior to tumor resection. The test dose was based on patient weight and stature and was a median of 1.25 mg (range 0.625 to 2.5 mg) for the study. Based on tumor visualization of the test dose, the authors calibrated the second dose accordingly and reported a median second dose of 1.875 mg (range 0.625 to 5 mg). With this dosing regimen, differential fluorescence was achieved in 82% of tumors; after excluding five oncocytomas that fluoresced as expected, 88% of tumors exhibited the desired fluorescence.

Table 1 includes a complete summary of all institutional studies utilizing ICG for RPN.

Table 1. Published institutional studies reporting outcomes of robotic partial nephrectomy (RPN) utilizing indocyanine green (ICG).

Study	Year	Institution	Patients (n)	Median Age (years)	Median Size (cm)	Median Operative Time (min)	Median WIT (min)	% Malignant	% Positive Margin
Tobis et al. [4]	2011	University of Rochester	9	69	3.8	181	19	91	0
Borofsky et al. [10]	2012	WFU, USC, NYU	27	60 *	2.8 *	256 *	0	81	0
Krane et al. [8]	2012	WFU	47	60 *	2.7	NR	15	79	6.4
Harke et al. [11]	2013	Missionsaerztliche, Klinik, Germany	22	63 *	3.8 *	156 *	12	50	0
Bjurlin et al. [12]	2013	NYU	48	54	2.6	155	17	75	3.8
Angell et al. [9]	2013	Ohio State University	79	55 *	3.5 *	187 *	13	79	0

* Mean values; WIT: warm ischemia time; NR: not reported; WFU: Wake Forest University; USC: University of Southern California; NYU: New York University.

3. Limitations of ICG

Although ICG has improved the visualization of renal vasculature and exophytic cortical tumors, this has not resulted in improved positive margin rates [8]. Perhaps this is secondary to an exophytic tumor being less likely to have a positive margin for experienced robotic surgeons. Furthermore, delineation of tumor margins for endophytic tumors has not improved with ICG [8]. Tobis *et al.* [4] hypothesized that poor visualization with endophytic tumors may be secondary to the normal cap of parenchyma over the tumor. However, they reported that decreased fluorescence could be seen once resection commenced and suggested that tumor margin identification for endophytic tumors may be most useful once normal parenchyma has been incised.

In our practice, we have identified other limitations for the use of ICG in RPN. First, the lack of uptake in peripelvic/intrarenal fat and the collecting system itself may prevent the ICG delineation of the most medial/central extent of the tumor. Second, when toggling between white light and near infrared fluorescence imaging, the surrounding intracorporeal field is dark in relation to the ICG illuminated kidney. Presumably, this may increase the risk of iatrogenic injury to surrounding structures for the robotic surgeon. Third, we have found that the green color of ICG dye may "bleed" when excising the mass, making the margins less clear when cutting. Fourth, anecdotally in our experience, ICG is not helpful for reoperative surgery, particularly during recurrence after laparoscopic partial nephrectomy and cryoablation, secondary to the renal scar and associated post-inflammatory changes hypoperfusing similar to the tumor. Finally, the continued need for frozen sections to confirm negative margins in challenging cases has not been avoided by the introduction of ICG to RPN.

4. Future Directions

One of the most significant benefits of using ICG is to confirm that the region of potential excision is truly ischemic; there may be instances where a bulldog clamp is placed on the renal artery however ICG identifies an additional artery(ies). Furthermore, with the ability of ICG near infrared fluorescent imaging to delineate the vascular anatomy, attention has been focused on improving zero-ischemia and selective clamping using ICG. Borofsky *et al.* [10] reported their experience of super-selective arterial clamping during zero-ischemia RPN in 27 patients compared to 27 patients undergoing conventional clamping RPN. Between the two cohorts there were no differences in baseline patient or tumor characteristics. Operative outcomes were comparable between the two groups, however the conventional renal artery-clamping group had a shorter operative time (212 *vs.* 256 min, $p = 0.02$). Post-operatively, the zero-ischemia group had superior kidney function compared to the conventional RPN group (reduction of estimated GFR -1.8% *vs.* -14.9%, $p = 0.03$). Both groups had comparable post-operative complication rates and all surgical margins were negative. Harke *et al.* [11] performed a single-surgeon matched-pair study of 22 patients who underwent selective clamping (tertiary vessels feeding the tumor) compared to 15 patients who underwent main renal artery clamping after administering ICG to both groups of patients. Between the groups there were no differences in demographic or perioperative data as well as post-operative complications. Short-term change in renal function was improved in the selective clamping group despite comparable baseline eGFR (6.2% *vs.* 18.6%, $p = 0.045$). These studies show that ICG-assisted vascular identification specific to the tumor may allow either off clamp or selective clamping of the arterial supply, demonstrating short-term post-operative improvement in renal function.

In an effort to assess whether ICG can predict malignancy in RPN patients, Manny *et al.* [13] reviewed the ICG fluorescence pattern in 100 patients undergoing RPN and correlated these findings to final tumor histology. The tumor fluorescent schema used included isofluorescent (the same amount as surrounding parenchyma), hypofluorescent (less than surrounding parenchyma, but with uptake), or afluorescent (no visible uptake of dye). Using a single intravenous dose of 5–7.5 mg of ICG before vascular clamping, 86 solid lesions were categorized as isofluorescent ($n = 3$, two clear cell and one translocation tumor) and hypofluorescent ($n = 83$, 65 malignant and 18 benign lesions). For determining malignant *vs.* benign lesions, hypofluorescence had a positive predictive value of 87%, negative predictive value of 52%, sensitivity of 84% and specificity of 57%. Given these relatively poor predictive tools, the authors determined that the role of predicting malignancy based on ICG fluorescent patterns remains to be determined and should not supplant preoperative imaging. Perhaps standardization of the ICG dosing regimen, as suggested by Angell *et al.* [9], may improve intraoperative prediction of malignancy and possibly guide the surgeon's aggressiveness during tumor resection.

Other compounds and dyes have been analyzed to assess differences in normal parenchyma and renal masses. For example, photodynamic diagnostic (PDD) has been used over the last decade to differentiate healthy from diseased tissue. Specifically, 5-aminolevulinic acid (5-ALA) accumulates in malignant cells and after endogenous metabolism to protoporphyrin IX can be irradiated at a wavelength of 390 to 440 nm and subsequently fluoresce red at 635 nm [14,15]. Hoda *et al.* [16] utilized an oral dose of 1.5 g of 5-ALA four hours prior to surgery in 77 patients undergoing laparoscopic partial nephrectomy and reported that 58 of 61 patients with renal cell carcinoma demonstrated a positive response to excitation light, however the remaining benign lesions also demonstrated some response to excitation light. Among the 61 malignant lesions, two patients had positive margins, however PDD was able to detect remaining tissue in the partial nephrectomy bed, allowing subsequent complete resection. To date, this technology applied to RPN has not been reported. Disciplines other than Urology have used additional fluorescent compounds to demonstrate combination nerve and tumor imaging for breast cancer surgery [17]. Furthermore, animal models have demonstrated the feasibility of robotic-assisted sentinel lymph node mapping for pre-sacral nodes of the prostate using Ga-68-labeled tilmanocept [18].

5. Conclusions

The initial institutional studies using near infrared fluorescent imaging of ICG dye demonstrate safety and feasibility when performing RPN. Identification of the renal vasculature and tumor margins is possible for mainly exophytic tumors; however, positive margin status for these tumors has not substantially improved. ICG may improve the ability of experienced surgeons to perform RPN with selective clamping or in some instances zero-ischemia, and early post-operative kidney function may improve using this technique. Although there are limitations to ICG for RPN, including visualization of endophytic tumors and difficultly demarcating tumor margins in reoperative cases, the field continues to advance as robotic surgeons become accustomed to ICG and additional compounds are developed. The precise combination of fluorescent compound, dose, and optimal tumor anatomy for ICG RPN has yet to be elucidated.

Author Contributions

Zachary Klaassen: Conceptual design, drafting of the manuscript, critical revision of the manuscript. Qiang Li: Critical revision of the manuscript. Rabii Madi: Critical revision of the manuscript. Martha K. Terris: Conceptual design, drafting of the manuscript, critical revision of the manuscript.

Conflicts of Interest

The authors declare no conflict of interest.

References

1. Huang, W.C.; Elkin, E.B.; Levey, A.S.; Jang, T.L.; Russo, P. Partial nephrectomy versus radical nephrectomy in patients with small renal tumors—Is there a difference in mortality and cardiovascular outcomes? *J. Urol.* **2009**, *181*, 55–61.
2. Yossepowitch, O.; Thompson, R.H.; Leibovich, B.C.; Eggener, S.E.; Pettus, J.A.; Kwon, E.D.; Herr, H.W.; Blute, M.L.; Russo, P. Positive surgical margins at partial nephrectomy: Predictors and oncological outcomes. *J. Urol.* **2008**, *179*, 2158–2163.
3. Secil, M.; Elibol, C.; Aslan, G.; Kefi, A.; Obuz, F.; Tuna, B.; Yorukoglu, K. Role of intraoperative US in the decision for radical or partial nephrectomy. *Radiology* **2011**, *258*, 283–290.

4. Tobis, S.; Knopf, J.; Silvers, C.; Yao, J.; Rashid, H.; Wu, G.; Golijanin, D. Near Infrared Fluorescence Imaging with Robotic Assisted Laparoscopic Partial Nephrectomy: Initial Clinical Experience for Renal Cortical Tumors. *J. Urol.* **2011**, *186*, 47–52.

5. Unno, N.; Suzuki, M.; Yamamoto, N.; Inuzuka, K.; Sagara, D.; Nishiyama, M.; Tanaka, H.; Konno, H. Indocyanine green fluorescence angiography for intraoperative assessment of blood flow: A feasibility study. *Eur. J. Vasc. Endovasc. Surg.* **2008**, *35*, 205–207.

6. Golijanin, D.J.; Marshall, J.; Cardin, A. Bilitranslocase (BTL) is immunolocalised in proximal and distal renal tubules and absent in renal cortical tumors accurately corresponding to intraoperative near infrared fluorescence (NIRF) expression of renal cortical tumors using intravenous indocyanine green (ICG). *J. Urol. Suppl* **2008**, *179*, 137.

7. Tobis, S.; Knopf, J.K.; Silvers, C.R.; Marshall, J.; Cardin, A.; Wood, R.W.; Reeder, J.E.; Erturk, E.; Madeb, R.; Yao, J.; *et al.* Near Infrared Fluorescence Imaging After Intravenous Indocyanine Green: Initial Clinical Experience with Open Partial Nephrectomy for Renal Cortical Tumors. *Urology* **2012**, *79*, 958–964.

8. Krane, L.S.; Manny, T.B.; Hemal, A.K. Is Near Infrared Fluorescence Imaging Using Indocyanine Green Dye Useful in Robotic Partial Nephrectomy: A Prospective Comparative Study of 94 Patients. *Urology* **2012**, *80*, 110–118.

9. Angell, J.E.; Khemees, T.A.; Abaza, R. Optimization of Near Infrared Fluorescence Tumor Localization during Robotic Partial Nephrectomy. *J. Urol.* **2013**, *190*, 1668–1673.

10. Borofsky, M.S.; Gill, I.S.; Hemal, A.K.; Marien, T.P.; Jayaratna, I.; Krane, L.S.; Stifelman, M.D. Near-infrared fluorescence imaging to facilitate super-selective arterial clamping during zero-ischemia robotic partial nephrectomy. *BJU Int.* **2012**, *111*, 604–610.

11. Harke, N.; Schoen, G.; Schiefelbein, F.; Heinrich, E. Selective clamping under the usage of near-infrared fluorescence imaging with indocyanine green in robot-assisted partial nephrectomy: A single-surgeon matched-pair study. *World J. Urol.* **2013**, doi:10.1007/s00345-013-1202-4.

12. Bjurlin, M.A.; Gan, M.; McClintock, T.R.; Volpe, A.; Borofsky, M.S.; Mottrie, A.; Stifelman, M.D. Near-infrared Fluorescence Imaging: Emerging Applications in Robotic Upper Urinary Tract Surgery. *Eur. Urol.* **2014**, *65*, 793–801.

13. Manny, T.B.; Krane, L.S.; Hemal, A.K. Indocyanine Green Cannot Predict Malignancy in Partial Nephrectomy: Histopathologic Correlation with Fluorescence Pattern in 100 Patients. *J. Endourol.* **2013**, *27*, 918–921.

14. Uehlinger, P.; Zellweger, M.; Wagnieres, G.; Juillerat-Jeanneret, L.; van den Bergh, H.; Lange, N. 5-aminolevulinic acid and its derivatives: Physical chemical properties and proporphyrin IX formation in cultured cells. *J. Photochem. Photobiol.* **2000**, *54*, 72–80.

15. Steinbach, P.; Weingandt, H.; Baumgartner, R.; Kriegmair, M.; Hofstadter, F.; Knuchel, R. Cellular fluorescence of the endogenous photosensitizer protoporphyrin IX following exposure to 5-aminolevulinic acid. *Photochem. Photobiol.* **1995**, *62*, 887–895.

16. Hoda, M.R.; Popken, G. Surgical Outcomes of Fluorescence-Guided Laparoscopic Partial Nephrectomy Using 5-Aminolevulinic Acid-Induced Protoporphyrin IX. *J. Surg. Res.* **2009**, *154*, 220–225.

17. Nguyen, Q.T.; Olson, E.S.; Aguilera, T.A.; Jiang, T.; Scadeng, M.; Ellies, L.G.; Tsien, R.Y. Surgery with molecular fluorescence imaging using activatable cell-penetrating peptides decreases residual cancer and improves survival. *Proc. Natl. Acad. Sci. USA.* **2010**, *107*, 4317–4322.

18. Stroup, S.P.; Kane, C.J.; Farchshchi-Heydari, S.; James, C.M.; Davis, C.H.; Wallace, A.M.; Hoh, C.K.; Vera, D.R. Preoperative sentinel lymph node mapping of the prostate using PET/CT fusion imaging and Ga-68-labeled tilmanocept in an animal model. *Clin. Exp. Metastasis* **2012**, *29*, 673–680.

3

Drive the Drive: From Discrete Motion Plans to Smooth Drivable Trajectories

Henrik Andreasson *, Jari Saarinen , Marcello Cirillo , Todor Stoyanov and Achim J. Lilienthal

Centre of Applied Autonomous Sensor Systems (AASS), Örebro University, 70182 Örebro, Sweden;
E-Mails: jari.saarinen@oru.se (J.S.); marcello.cirillo@oru.se (M.C.); todor.stoyanov@oru.se (T.S.);
achim.lilienthal@oru.se (A.L.)

* Author to whom correspondence should be addressed; E-Mail: henrik.andreasson@oru.se

External Editor: Huosheng Hu

Abstract: Autonomous navigation in real-world industrial environments is a challenging task in many respects. One of the key open challenges is fast planning and execution of trajectories to reach arbitrary target positions and orientations with high accuracy and precision, while taking into account non-holonomic vehicle constraints. In recent years, lattice-based motion planners have been successfully used to generate kinematically and kinodynamically feasible motions for non-holonomic vehicles. However, the discretized nature of these algorithms induces discontinuities in both state and control space of the obtained trajectories, resulting in a mismatch between the achieved and the target end pose of the vehicle. As endpose accuracy is critical for the successful loading and unloading of cargo in typical industrial applications, automatically planned paths have not been widely adopted in commercial AGV systems. The main contribution of this paper is a path smoothing approach, which builds on the output of a lattice-based motion planner to generate smooth drivable trajectories for non-holonomic industrial vehicles. The proposed approach is evaluated in several industrially relevant scenarios and found to be both fast (less than 2 s per vehicle trajectory) and accurate (end-point pose errors below 0.01 m in translation and 0.005 radians in orientation).

Keywords: motion planning; motion and path planning; autonomous navigation

---

1. Introduction

Automatically Guided Vehicles (AGVs) have been deployed in large numbers for industrial intra-logistic tasks. They typically transport payload from a loading to a drop-off location and it is crucial that they drive accurately to given poses. Being able to *arrive at a defined pose with high accuracy and precision* is a fundamental requirement for AGV systems. It is especially important since most AGV platforms, e.g., forklift trucks [1] or waist-actuated wheel loaders [2], are non-holonomic. Thus, even in cases when an error in the final pose can be reliably detected, it is not possible for the vehicle controller to correct it over a short distance. According to the AGV system provider Kollmorgen [3], the required end pose accuracy for picking up pallets is 0.03 m in position and 1 degree (0.017 radians) in orientation, for example.

In current commercial AGV solutions, all paths need to be defined manually before operation. This is a time-consuming, inflexible and costly procedure, which must be repeated for every new deployment or whenever the area of operation changes. An even more important disadvantage is that paths cannot be changed during operation to respond to obstacles. In order to enable new applications, it is key to replace this inflexible off-line process with on-line motion planning. State-of-the-art motion planners (based either on Rapidly Exploring Random Trees (RRT), Probabilistic Roadmaps (PRM) or state space lattice search) have so far not been demonstrated to generate trajectories that enable AGVs to achieve the required goal pose accuracy for industrial applications. In addition, the trajectories obtained by current planners are not guaranteed to be directly drivable—discontinuities in vehicle controls and within the trajectories are often left untreated and have to be handled by the controller. This is a well known problem which has recently received attention [4,5], but, to-date, no solution has been proposed, which can both guarantee the required accuracy and be used on-line.

The focus of this paper is on on-line motion planning for car-like vehicles given arbitrary end poses. We propose and evaluate a path smoothing approach to improve end pose accuracy obtained with paths generated by a lattice-based motion planner. Of course, the achieved end pose accuracy depends on all navigation subsystems, including motion planning, localization and motion execution. Therefore, we developed a complete navigation system, composed of a lattice-based motion planner, the continuous space path smoother proposed in this paper, a trajectory generator and a model predictive controller. We compare our system against a state-of-the-art commercial AGV solution. For the comparisons in this paper, we rely on a commercial reflector-based localization module.

The contribution of this paper is two-fold. First, we propose a new path smoothing approach that can be used in a complete, on-line navigation system to produce highly accurate motions for car-like vehicles. Our approach does not require a transformation of the input path to a new representation, which is smooth by definition (e.g., splines). Thus, it can directly incorporate state-space constraints in the smoothing process, avoiding the need to verify the path again after smoothing. In our experiments, we show that our approach generates smooth drivable trajectories for arbitrary goal poses in under 2 s (using a single threaded implementation with an i7-2860QM CPU at 2.50 GHz), a fraction of the runtime reported by previous approaches [4]. Second, we describe our complete navigation system and present an extensive experimental evaluation where a non-holonomic vehicle has to sequentially drive to a number of given goal poses (drawn as black arrows in the example depicted in Figure 1). We show the importance of the path smoother and demonstrate that our system can generate and execute smooth paths (see the example in Figure 1 containing paths obtained from a lattice-based motion planner, the corresponding smoothed paths and the actual trajectories driven by the vehicle) with end pose errors comparable to a commercial AGV system based on manually defined paths.

Figure 1. Example of the problem addressed in this paper.

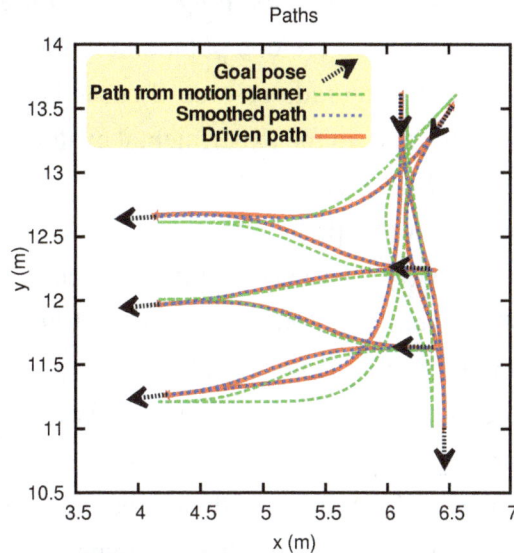

In Section 2 we present related work on motion planning. We detail our system in Section 3 and present the experimental evaluation in Section 4. Finally, Section 5 concludes with a discussion and presents directions for future research.

2. Related Work

The industry standard for autonomous navigation is to use predefined trajectories that the AGVs follow strictly. The trajectories are either manually defined or learned through teaching-by-demonstration from a human operator [6,7]. Although conceptually simple, this approach has drawbacks such as high deployment costs and lack of flexibility. A change in the configuration of a warehouse or an additional loading point, for instance, require the manual definition of a new set of trajectories. In addition, if an AGV encounters an unforeseen obstacle during operation, it can only employ very simple strategies (typically stopping until the obstacle moves or is removed). To overcome these drawbacks, many different techniques for automatic path and trajectory generation have been proposed in the past decades.

Combinatorial methods are not very well suited in the presence of differential constraints (e.g., kinematic constraints for non-holonomic vehicles) and analytical solutions cannot effectively cope with obstacles [8]. To overcome these problems, sampling-based approaches have been introduced and studied in recent years. In particular, three families of methods are currently widely used: Probabilistic Roadmaps (PRMs) [9], Rapidly-exploring Random Trees (RRTs) [10–12] and lattice-based motion planners [13,14]. All sampling-based approaches have been shown to be effective in high-dimensional configuration spaces. Lattice-based motion planners, in particular, combine the strengths of the approaches discussed above with well studied classical AI graph exploration algorithms, such as A^*,

ARA^* and D^*Lite [15]. Differential constraints are incorporated in the search space by means of pre-computed motion primitives, which sample the state space on a regular lattice. The search space is then explored using efficient graph search techniques. However, all sampling-based approaches to non-holonomic motion planning generate paths and trajectories that typically present discontinuities, within the trajectory itself or between the terminal and goal state. This is a known problem and prevents current solutions to achieve the end pose accuracy required by industrial applications. In recent years, several solutions have been suggested for smoothing the trajectories from sampling-based motion planners [4,5]. So far, the methods suggested are computationally too expensive to be used on-line, as smoothing requires time in the order of hundreds or even thousands of seconds [4]. By contrast, our approach is designed to be used on-line and, in the experimental evaluation, we show that the smoothing step usually takes less than two seconds (see Table 1).

A different approach to obtain smooth paths is to start with a set of waypoints, and then to switch to a representation that inherently guarantees continuous curvatures. Popular algorithms take as input a set of waypoints, obtained either automatically [16,17] or by manually driving the vehicle on the desired path [6,7]. Then an optimization procedure is employed, which operates directly on the parameters of the new representation rather than on the waypoints. Common representations are Quintic splines [18], B-splines [17] or clothoids [19,20]. Fitting the new representation to the waypoints in general entails a change of the original path and therefore does not guarantee collision-free paths. Thus, a post-processing step is necessary to check whether the path in the new representation is still collision-free [16,21]. Non-holonomic vehicles have additional constraints on the curvature of the motion, *i.e.*, the maximum steering angle of the vehicle. To guarantee drivable paths, also these constraints require a time-consuming verification after path smoothing by transformation into a different representation.

3. Generating Smooth Trajectories On-Line

We developed our approach to path smoothing and to autonomous navigation to fulfill the stringent requirements for autonomous vehicles in industrial environments. For the definition of these requirements, we relied on the long experience of our industrial partner Kollmorgen, a world leader in providing AGV solutions. Kollmorgen has deployed approximate 15,000 AGVs since 1991 in different industrial settings. According to Kollmorgen, high end pose accuracy is of paramount importance. For safe loading and unloading of pallets, for example, the required end pose accuracy is 0.03 m in position and 1 degree (0.017 radians) in orientation.

In this work, we consider trajectories for AGVs which operate as a fleet for warehouse automation. At the core of the system is a central vehicle coordinator [22], which can directly control the speed of each autonomous vehicle. Thus, the paths and speed profiles of the controlled vehicles have to be computed separately, in order to allow the coordinator to prevent deadlocks. This is a desirable separation whenever multiple AGVs need to be coordinated. For our system, we adopt the three-step approach shown in Figure 2. First, we calculate kinematically drivable paths with a lattice-based motion planner for each vehicle. Then, the path is processed by the path smoother proposed in this paper, resulting in a continuous drivable path. Finally, a trajectory generator associates speed profiles consistent with dynamic and coordination constraints to the smoothed paths and generates the final trajectories that the vehicle controller executes [23].

Figure 2.   Overview of the processing steps detailed in Sections 3.2–3.4 and their connections.

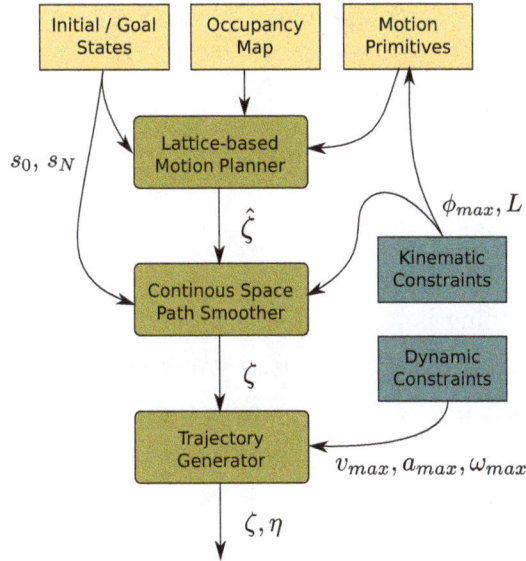

3.1. Problem Formulation

In this paper we consider car-like vehicles (standard trucks, forklifts, *etc.*) that are commonly found in indoor production sites. The control space $u = (v, \omega)$ of such vehicles is composed of forward speed v and steering velocity ω. The state space $s = (x, y, \theta, \phi)$ consists of all vehicle configurations composed of 2D position (x, y), heading θ and steering angle ϕ. Let ζ and η be sets of N points in state space and control space respectively, defined as:

$$\begin{aligned} \zeta = \{s_i\} &= \{(x_i, y_i, \theta_i, \phi_i)\} \\ \eta = \{u_i\} &= \quad \{(v_i, \omega_i)\} \end{aligned} \tag{1}$$

We assume a fixed time step $\Delta T = \frac{T}{N} (= 60\ ms)$, where T is the time for reaching the goal. The state transition of car-like vehicles $\dot{s} = f(s, u, t)$ is computed as follows:

$$\begin{aligned} \dot{x}_i &= v_i \cos \theta_i \\ \dot{y}_i &= v_i \sin \theta_i \\ \dot{\theta}_i &= v_i \frac{\tan \phi_i}{L} \\ \dot{\phi}_i &= \omega_i \end{aligned} \tag{2}$$

where L is the distance between front and back axles.

Given an initial state $\bar{s}_0 = (x_0, y_0, \theta_0, \phi_0)$ and a goal state $\bar{s}_N = (x_N, y_N, \theta_N, \phi_N)$ for a vehicle, we want to calculate a set of N control points η and the corresponding state points ζ, which can take the vehicle from $\bar{s}_0$ to $\bar{s}_N$. A solution to the problem is valid if also additional vehicle-dependent constraints are respected, such as the maximum allowed steering angle ϕ_{max}, the maximum steering velocity ω_{max}, the maximum forward and reverse velocity v_{max} and the maximum acceleration $|\dot{v}_i| \leq a_{max}$. In our experiments, the x_0, y_0 and θ_0 components of the start state s_0 are obtained directly from an off-the-shelf localization system, while the steering angle ϕ_0 is taken from absolute encoder readings. The goal state is always assumed to have a steering angle component $\phi_N = 0$. See also the overview depicted in Figure 2.

3.2. Lattice-Based Motion Planner

The first step of our approach is a lattice-based motion planner [24], which quickly computes kinematically feasible paths, optimized with respect to a cost function that considers distance traveled and penalizes backwards and turning motions. Given a model of vehicle maneuverability, the intuition behind lattice-based motion planning is to sample the state space in a regular fashion and to constrain the motions of the vehicle to a lattice graph, that is, a graph embedded in a Euclidean space $\mathbb{R}^n$ which forms a regular tiling [25]. Each vertex of the graph represents a valid configuration of the vehicle, while each edge encodes a motion which respects the non-holonomic constraints of the vehicle. A valid configuration for a vehicle is a four-dimensional vector $c = \langle x, y, \theta, \phi \rangle$, where (x, y) lies on a grid of resolution r, $\theta \in \Theta$ and $\phi \in \Phi$. In the experiments presented in this paper, r is equal to 0.2 m, $|\Theta| = 16$ and $|\Phi| = 1$. In particular, Θ is the set of all the angles in $[0, 2\pi)$ which are multiples of $\frac{\pi}{8}$ and $\Phi = \{0\}$, which means that we only consider configurations where the steering angle is equal to 0 with respect to the vehicle itself. This is because reducing the cardinality of Φ reduces the search space and solutions can be calculated quickly. The fact that at this stage the steering angle of the vehicle is assumed to be equal to 0 at the beginning and at the end of every motion is then compensated by our smoother (for the computation times of the planner in our experiments, see Table 1, "Motion planning").

Table 1. Computational time for motion planning and path smoothing (Section 4.3).

	Mean (s)	Std (s)	Max (s)
Motion planning	0.105	0.085	0.680
Path smoothing	1.095	0.222	1.868

The planner uses a set of pre-computed, kinematically feasible motion primitives, which are repeatedly applied to obtain a directed graph which covers the configuration space. Information about the static obstacles in the environment is provided to the planner by an occupancy map and is used to prune the search graph to obtain collision-free paths. The motion primitives are automatically generated to fully capture the mobility of the vehicle and then reduced for efficiency purposes, as described in [26], without compromising the reachability of the configuration space of the vehicle. The graph is then explored using A^*, or ARA^* [27], one of its most efficient anytime versions, which can provide provable bounds on sub-optimality. Effective heuristic functions [28] and pre-computed vehicle footprints of each motion primitive for fast collision detection are employed to speed up the exploration of the lattice.

Given a start and a goal state (s_0, s_N), the motion planner generates an obstacle free, kinematically drivable path $\hat{\zeta}$, which means that the steering angle constraint $-\phi_{max} \leq \phi \leq \phi_{max}$ (Equation (6)) is respected. Note that the planner at this stage does not generate the controls for the vehicle. The path $\hat{\zeta}$, however, brings the vehicle from a discretized start state $\hat{s}_0$ to a discretized goal state $\hat{s}_N$, where $\hat{s}_0$ and $\hat{s}_N$ represent the closest states on the lattice to s_0 and s_N, respectively. This introduces an error on the order of the lattice discretization. Moreover, $\hat{\zeta}$ is by construction guaranteed to be drivable by the vehicle at a nominal speed and it is C^1-continuous, but not necessarily C^2-continuous.

3.3. Continuous Space Path Smoother

The path $\hat{\zeta}$ obtained by the motion planner in the previous section has three distinct problems, which prevent us from directly feeding it to the vehicle controller. First, the start and goal states used by the planner are discretized and do not necessarily correspond to the given initial and end state (see Figure 1). Second, the planner assumes that the steering angle at the start state is always equal to zero, which may not correspond to the actual state of the vehicle. Third, $\hat{\zeta}$ is not necessarily C^2-continuous, which means that the rate of change of the steering angle of the vehicle can be discontinuous (see Figure 3). While some of these issues can be handled and corrected by the controller, the accuracy of the final vehicle pose with respect to the target pose would be at best on the order of the grid discretization of the planner (as demonstrated in the experiments in Section 4). In order to solve these problems and improve navigation, our approach makes small local modifications to $\hat{\zeta}$.

Figure 3. Comparison between the original path from the lattice-based motion planner and the smoothed path after applying our continuous space path smoother. (a) paths; (b) steering angles.

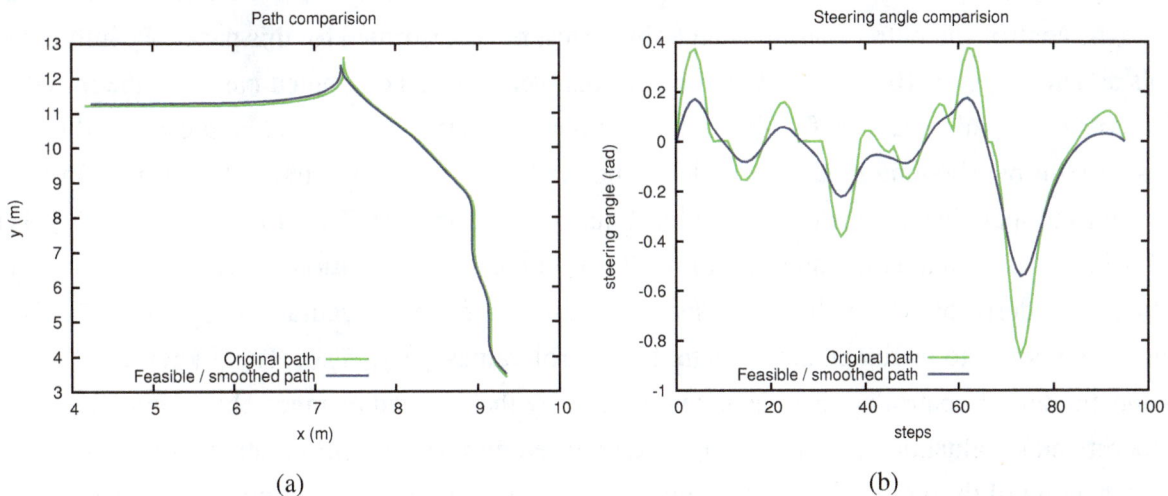

(a)

(b)

We begin by formulating the above problems as constraints on the states of $\hat{\zeta}$ so as to obtain a set of states and the corresponding control inputs $\{s_i, u_i\}_{i=0...N}$ that satisfy them. First, the start and end states of the modified path need to correspond precisely to the current vehicle state $\bar{s}_0$ and the actual goal state $\bar{s}_N$:

$$\bar{s}_0 - s_0 = 0 \tag{3}$$

$$\bar{s}_N - s_N = 0 \tag{4}$$

These constraints address inaccuracies due to discretization and the assumption of a zero initial steering angle. The additional constraints in Equations (5)–(9) make sure that the path is conform with the kinematic constraints of the vehicle and the constraints on vehicle controls:

$$s_{i+1} = f(s_i, u_i), \ i = 1, \ldots, N-1 \tag{5}$$

$$\phi_{min} \leq \phi \leq \phi_{max} \tag{6}$$

$$-v_{max} \leq v \leq v_{max} \tag{7}$$

$$-\omega_{max} \leq \omega \leq \omega_{max} \tag{8}$$

$$-a_{max} \leq a \leq a_{max} \tag{9}$$

The constraints in Equations (3)–(9) define the set of all possible executable trajectories to bring the vehicle from a given start to a goal state without considering obstacles. An uninformed search through this set of feasible trajectories is a very inefficient way of solving the continuous space motion planning problem in the presence of obstacles. The key to the efficiency that we obtain with the proposed approach is that an uninformed search is not necessary. From the lattice-based motion planner we already have a cost-optimal, obstacle-free path $\hat{\zeta}$. Although this path generally lies outside the feasible set defined by Equations (3)–(9) we can use it to define a boundary value optimization problem in which we initialize the variables $\{s_i\}_{i=0\ldots N}$ with the path $\hat{\zeta}$. We then perform a search for the closest trajectory in the feasible set using a direct multiple shooting method implementation from the ACADO Toolkit [29].

The multiple shooting method [30] assumes that the controls u_i are discretized piecewise over time and kept constant during each step ΔT. It utilizes the ordinary differential equations (ODEs) defined above (Equation (2)) to integrate the control value u_i over the time ΔT from an initial state value s'_i to a final state value s'_{i+1}. In order to obtain a smoothed path, and not only a feasible one, it is important to initialize the control values to a constant value (in the test runs performed for this paper, we initialized the controls as $\{u_i\}_{i=0\ldots N} = (0,0)$). If the initial controls were instead computed based on the initial given path, the obtained path would be feasible but not smoothed, as the turns would be preserved through the process. A multiple shooting method is divided into the following steps: First, a Non-Linear Program is defined so as to include the constraints listed in Equations (5) and (6). The Non-Linear Program is then solved using a Sequential Quadratic Program (SQP). In each SQP iteration the ODEs are recomputed, including their derivatives, which are subsequently used to form a Quadratic Program (QP). The QP is then solved, so as to incrementally update the control values $\{u_i\}_{i=0\ldots N}$. This incremental approach limits the amount of control applied, as it can be seen by the reduced changes observable in steering in our experimental evaluation (see Figure 3b). The required number of SQP iterations was typically 3.

One key point of this work is that we formulate the problem as a standard optimization problem. This choice allows us to directly benefit from using the vast amount of freely available optimization tools. We also tried to solve the problem using the single shooting method implementation in ACADO Toolkit [29], and this yielded similar results. Our approach is largely agnostic to a change of the optimization tool since we do not minimize any specific objective function. All the tests presented in this paper, however, were performed using the multiple shooting method.

The output of this optimization phase is a trajectory (ζ, η), including a set of vehicle controls. Since there is no objective which optimizes the speed of the vehicle, the returned control values are only guaranteed to fulfill the kinematic constraints of the vehicle. To obtain the fastest possible trajectory would require to solve a more complex problem, *i.e.*, an Optimal Control Problem, which would be too time consuming. The decoupling of the generation of the control set from the path is further motivated by the fact that the fastest trajectory profile may not necessarily be the optimal one in a multi-robot navigation scenario, as it is common in intra-logistics applications. Our full navigation system employs a centralized coordination scheme [23], whose description is outside the scope of this paper. For the purpose of experimentally evaluating our path smoothing approach, here we rely on a subsequent trajectory generation step (Section 3.4). Because of this decoupling, we can simplify the boundary value optimization problem by delegating the constraints regarding velocities and accelerations (Equations (7)–(9)) to the trajectory generator. In addition, we subsample the number of states and control points N to further reduce the computational time (in our experiments N was 100).

It is important to note that the path smoother changes the path from the motion planner and therefore we cannot guarantee that it is still collision-free. However, the deviation from the input path is very small (see Figure 3). This is because the smoother searches for the closest trajectory in the feasible set starting from the original path. In our current approach we compensate for this limited deviation by assuming an enlarged vehicle footprint during motion planning (width and length were both expanded by 10%). A rigorous way to solve this problem would be to specify boundary conditions on every state during the smoothing step. These constraints would guarantee that the new path is obstacle free. This aspect is left for future work.

3.4. Trajectory Generator

The trajectory generator takes as input the states computed by the path smoother and it works in a similar way as described in [16,31]. More specifically, it assigns the largest possible velocity to each state in the path within the given constraints. The constraints are the boundary conditions with initial and final velocities and limitations on steering velocity, speed and acceleration. The output of this module is a trajectory with a fixed ΔT of 60 ms, used by the controller. In the current implementation we use linear interpolation, which is effective since the distance between interpolation states is small.

In the experimental evaluation the parameters for the maximum velocity, acceleration and steering angle velocity were set to $v_{max} = 0.5 \ m/s$, $a_{max} = 0.2 \ m/s^2$ and $\omega_{max} = 1 \ rad/s$, respectively.

3.5. Model Predictive Controller

In our vehicle navigation system, we use a Model Predictive Controller (MPC) [23]. The core idea of this type of controllers is to model how the states of the vehicle evolve over a time preview window, given a set of control inputs. The controller then optimizes the control output using an objective function based on the trajectory to follow. In our implementation we use a preview window of 25 control steps and each control step has a duration of 60 ms. The controller gains were set to equally weight heading and distance. This is reflected in the results presented in Figure 4: the heading and distance errors are quantitatively very similar. As the controller is not the main focus of this paper, we use it with default parameters as a "black box" in the experimental evaluation.

Figure 4. Navigation pose accuracy. (**a**) side distance errors; (**b**) heading angle errors. The goal IDs refer to Figure 5b. Note that the paths generated by the planner and not processed by the smoother present errors one order of magnitude larger than the ones in the figure and therefore are omitted.

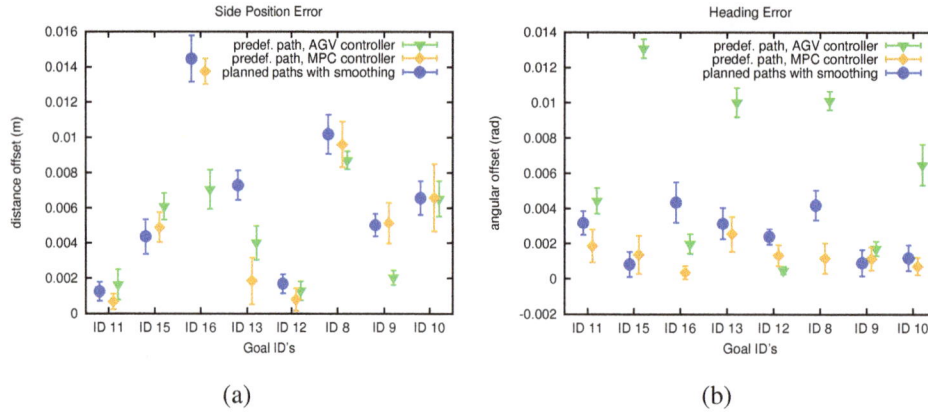

(a)   (b)

4. Experimental Evaluation

We evaluated our navigation system using an industrially-relevant vehicle configuration. We deployed an AGV operator training platform from Kollmorgen (Figure 5a) in a small-scale test environment at Örebro University. The vehicle kinematics are the same as those of standard fork lift trucks, with a combined steer and drive wheel. A commercial reflector-based localization solution using 22 reflective beacons was deployed in the test environment. This solution guarantees localization accuracy within less than 1 cm in translation and 0.001 radians in orientation. Using the on-board control system, we can access encoder and localization data and we can set steer and drive commands. All of the remaining components of our system run on a standard laptop with an i7-2860QM CPU at 2.50 GHz.

Figure 5. (**a**) AGV test platform used in the experiments; (**b**) Test layout for the comparison of the systems: the blue dots indicate goal poses identified by an ID. The turquoise line in the dots shows the goal headings.

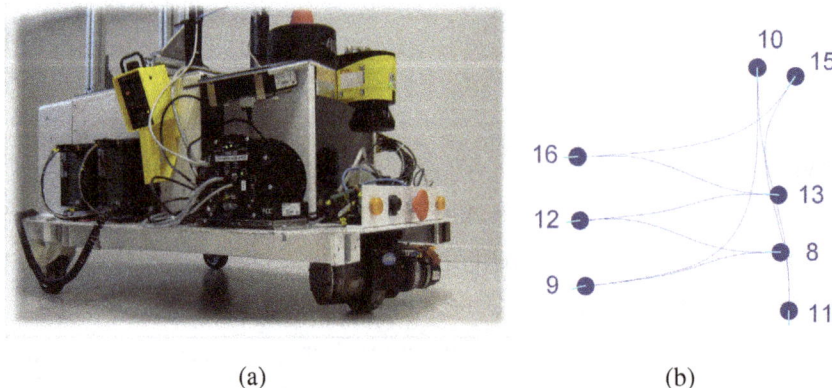

(a)   (b)

The overall goal of our experimental evaluation is to demonstrate the accuracy achieved while following the on-line generated trajectories and the reliability of our results. We structure our evaluation in several consecutive parts. First, we compare the controller of our system with the one of the commercial solution over identical, manually crafted trajectories. This set of tests is necessary to guarantee that the results obtained are comparable when using different approaches to path generation and are not biased by different performances at the controller level. The following evaluation step is the most crucial: once established that the two controllers have comparable capabilities, we extract the goal poses from the pre-defined paths and use them as input to our system. This means that we compare our approach, where the paths are automatically generated, with the performance of the commercial system which needs hand coded paths. These experimental runs allow us to demonstrate that our approach can entirely substitute the commercial system without loss of accuracy, thus rendering the time consuming procedure of manual path drawing obsolete. Finally, we test the robustness of our system over randomly generated goal poses and we evaluate the system over longer distances and analyze how the different numbers of control points N used in the path smoother affects the positioning accuracy.

4.1. Controller Comparison

We first created a set of B-spline parametrized trajectories T_{eval} (Figure 5b) with an AGV layout drawing tool provided by Kollmorgen. Special care was taken to make these trajectories as smooth as possible. We then used both controllers to follow the trajectories. Each trajectory was executed 10 times. The AGV controller can follow the given trajectories directly, whereas our controller extracts the paths P_{eval} from the layout trajectories T_{eval} and attaches speed profiles to them, as provided by our trajectory generator (Section 3.4). The tracking performance is shown in the first two lines of Table 2. In particular, we show the performance with respect to the final forward error, side error and heading error, as represented in Figure 6. We separated the errors because, for the controller of a non-holonomic vehicle, it is most difficult to control the heading and the position perpendicular to the direction of motion. For each type of error, we show the results in terms of mean, standard deviation and maximum value obtained in the test runs.

Table 2. Forward and side translation errors and orientation errors

Method	Forward Error (m)	Std (m)	Max (m)	Side Error (m)	Std (m)	Max (m)	Heading Error (rad)	Std (rad)	Max (rad)
predef. path, AGV controller	0.0168	0.0027	0.0224	0.0047	0.0028	0.0098	0.0060	0.0044	0.0138
predef. path, MPC controller	0.0025	0.0018	0.0080	0.0054	0.0044	0.0150	0.0013	0.0010	0.0041
planned path with smoothing	0.0018	0.0016	0.0080	0.0063	0.0042	0.0172	0.0025	0.0016	0.0062
planned path without smoothing	0.0273	0.0299	0.0886	0.0521	0.0211	0.1116	0.0621	0.0602	0.1783
60 random goals	0.0036	0.0036	0.0259	0.0084	0.0048	0.0231	0.0014	0.0013	0.0069

Figure 6. The error metrics used in the evaluations.

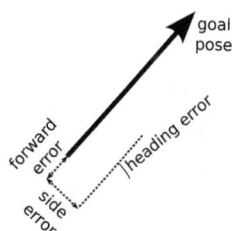

The commercial AGV controller has a higher forward error (statistically significant using unpaired t-test; $p < 0.0001$) which is partly due to an application-specific configuration compared to the MPC used in our system. The heading error is also higher (statistically significant; $p < 0.0001$) but on the other hand the side error is similar (difference not statistically significant; $p = 0.2317$). Both controllers are, however, fully capable of tracking the given trajectories.

4.2. Path Smoothing of Automatically Generated Paths

After we have established comparable performances of the two controllers, we now compare our complete navigation approach, where paths are automatically generated, with the commercial system, which needs hand coded paths.

We extracted the goal poses from the same paths P_{eval} employed in Section 4.1 and used them as targets for our lattice-based motion planner. The output of the planner is passed to the path smoother and then executed by the MPC controller. Table 2 (third and fourth rows) shows results obtained when the motion planner is used with or without path smoothing. Ideally, the automatically generated paths should allow for a comparable pose accuracy as the manually defined paths P_{eval}. This is indeed the case in our tests. Please note that in row 4, where the paths are not post-processed by the smoother, the required goal state is set as the last point in the trajectory, thus allowing the controller to correct the state error within the preview window. It can be seen that, in this case, the controller can compensate more effectively for errors in the direction of motion.

These results confirm that our system can produce smooth trajectories which allow very high end pose accuracy, both the heading and the forward error is improved (statistically significant using unpaired t-test; $p < 0.0001$), whereas the side error is slightly higher (t-test; $p = 0.0052$). Our approach can entirely substitute the commercial system without loss of accuracy, which makes the time-consuming procedure of manual path definition obsolete. Our results further show the necessity of path smoothing to obtain the required final pose accuracy, which is not possible with state-of-the-art motion planners that need to sample the continuous state space.

A key evaluation metric for AGV systems is their ability to repeatedly reach the same goal pose given a specific path. In an industrial scenario with manually predefined paths, precision is often more important than accuracy since a bias in the end pose can be compensated by adding the corresponding offset to the goal pose. For on-line motion planning, however, the key evaluation metric is end pose accuracy. In either case we are interested in a small variance over the final pose reached. The standard deviation is shown for different goals in Figure 4 and all end poses reached are shown for selected goals in Figure 7a,b. With path smoothing, the end pose variance is small and similar to the one obtained with the commercial system.

Figure 7. End pose accuracy on two different goal poses. (**a**) goal 10; (**b**) goal 13.

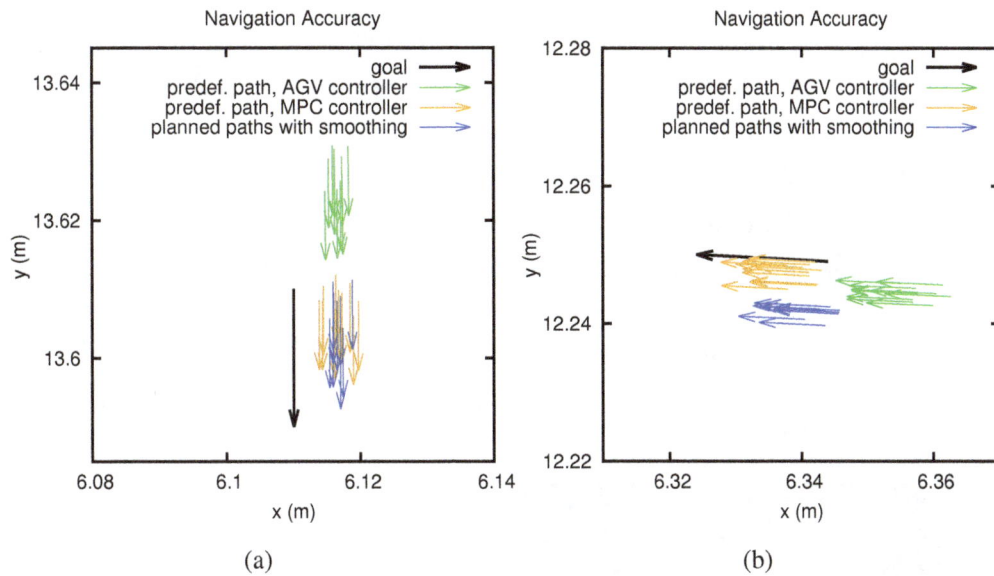

(a)   (b)

4.3. Evaluation over Randomly Chosen Goals

We further tested the robustness of our system with arbitrary goal poses and over longer distances, by generating a random set of 30 goal poses in two regions of the test environment, marked with dashed lines in Figure 8a. The vehicle traversed back and forth between poses, from one region to the other, for a total of 60 stops. Kernel density estimates of the position and heading errors are plotted in Figure 8b. Side, forward and heading errors are shown in the bottom row of Table 2, while computation times are presented in Table 1 (mean, max and standard deviation).

Figure 8. (**a**) Path driven while navigating to 60 random goals; (**b**) Kernel density estimate of corresponding end pose accuracy.

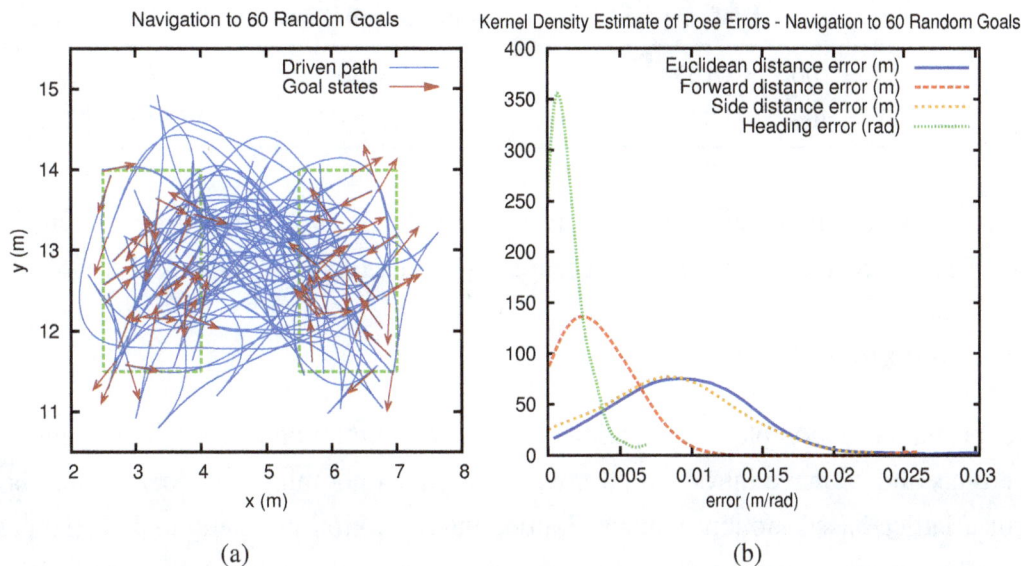

(a)   (b)

4.4. Evaluation of Discretization Effect on Longer Paths

In order to obtain a path quickly, the number of discretization steps needs to be kept low. However, if the step size is too big, the path smoother will not find paths of the same quality. To evaluate the impact of different numbers of discretization steps, we used again the goal poses shown in Figure 5b, but we interleaved these with an additional goal pose placed further away (see Figure 9a). We compared two different sets of test runs, where the number of discretization steps was set to either 100 or 400. The corresponding end pose accuracy and computation times are shown in Table 3.

Figure 9. (**a**) Paths and goal poses used in Section 4.4; (**b**) End pose accuracy of the bottom right goal state.

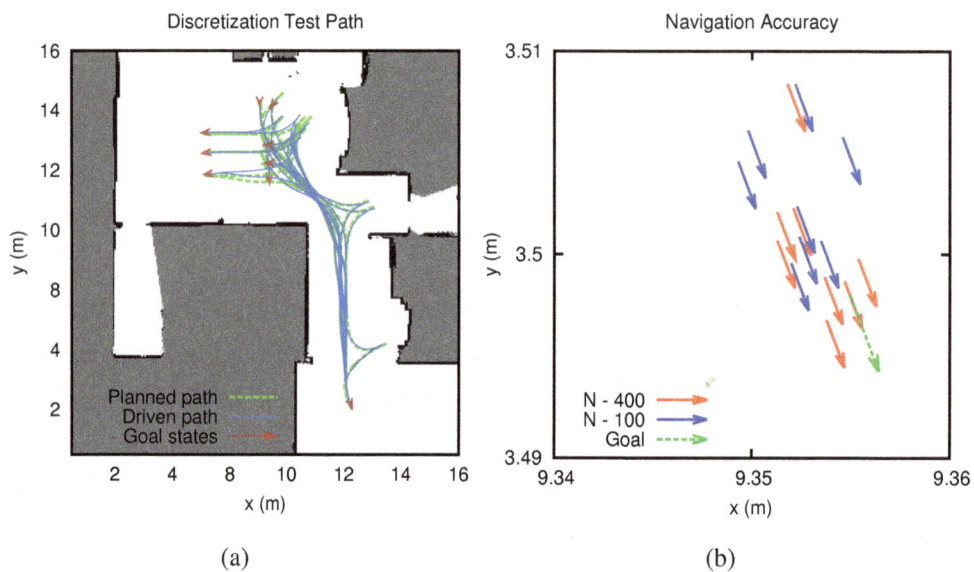

(a)   (b)

Table 3. End pose accuracy and computation time, different discretizations N.

	Dist (m)	Std (m)	Computation Time (s)	Std (s)
$N = 100$	0.0103	0.0091	1.5286	0.1668
$N = 400$	0.0079	0.0034	8.2057	1.0059

The results shows a minor improvement, but not-statistically significant (performing an unpaired t-test; $p = 0.4961$), in accuracy (as reported in Figure 9b) at the expense of increased computational costs.

5. Conclusions and Future Work

We have presented a complete navigation system for autonomous, non-holonomic vehicles in industrial settings and we have introduced a novel, fast path smoothing approach, which is applied to the output of a lattice-based motion planner. Lattice-based motion planners, as it is the case with all sampling-based approaches to motion planning, produce motions that present discontinuities which lead to insufficient accuracy and precision in industrial applications. This problem is overcome by our novel approach to path smoothing. Our approach has two major advantages: it can work on-line and uses the same state representation as the motion planner. It could thus directly incorporate state-space constraints in the smoothing process and therefore avoid the need to verify the path again after smoothing. This enables direct initialization of the path smoother with the automatically generated obstacle-free paths. Curvature checks on the final path are not necessary and we can directly include additional constraints in the optimization process. This opens up new possibilities to constrain the vehicle when it approaches the goal state. Constraining steering more in the last segments of the path could, for instance, improve pose accuracy even further. This, however, is left for future work. Another interesting avenue for further investigation will be to apply our approach to other planners, RRT- and PRM-based, to ascertain the generalizability of our methodology.

We have also presented an extensive experimental evaluation of our complete system and of all its major components. In the evaluation, we have compared our system with a state-of-the-art commercial solution, obtaining comparable results with respect to accuracy and precision. However, our system has the advantage that it can automatically plan trajectories on-line, instead of relying on expensive manual drawing.

Acknowledgments

This work was supported by the Swedish Knowledge Foundation (KKS) under projects: SAVIE, SAUNA and Semantic Robots.

Author Contributions

Henrik Andreasson worked on the main part of the experimental evaluation together with Jari Saarinen. The implementation was done by Henrik Andreasson with assistance of Marcello Cirillo and Todor Stoyanov. Henrik Andreasson, Jari Saarinen, Marcello Cirillo, Todor Stoyanov and Achim J. Lilienthal all contributed to the preparation of the manuscript. All listed authors discussed and approved the final manuscript.

Conflicts of Interest

The authors declare no conflict of interest.

References

1. Bouguerra, A.; Andreasson, H.; Lilienthal, A.J.; Åstrand, B.; Rögnvaldsson, T. MALTA: A System of Multiple Autonomous Trucks for Load Transportation. In Proceedings of the European Conference on Mobile Robots (ECMR), Mlini/Dubrovnik, Croatia, 23–25 September 2009.
2. Magnusson, M.; Almqvist, H. Consistent Pile-Shape Quantification for Autonomous Wheel Loaders. In Proceedings of the IEEE/RSJ International Conference on Intelligent Robots and Systems (IROS), San Francisco, CA, USA, 25–30 September 2011.
3. Kollmorgen web page. Available online: http: www.kollmorgen.com (accessed on 10 December 2014).

4. Cheng, P.; Frazzoli, E.; LaValle, S. Improving the performance of sampling-based motion planning with symmetry-based gap reduction. *IEEE Trans. Robot.* **2008**, *24*, 488–494.

5. Seiler, K.M.; Singh, S.P.; Sukkarieh, S.; Durrant-Whyte, H. Using Lie group symmetries for fast corrective motion planning. *Int. J. Robot. Res.* **2012**, *31*, 151–166.

6. Hellstrom, T.; Ringdahl, O. Follow the past: A path-tracking algorithm for autonomous vehicles. *Int. J. Veh. Auton. Syst.* **2006**, *4*, 216–224.

7. Marshall, J.; Barfoot, T.; Larsson, J. Autonomous underground tramming for center-articulated vehicles. *J. Field Robot.* **2008**, *25*, 400–421.

8. LaValle, S.M. *Planning Algorithms*; Cambridge University Press: Cambridge, UK, 2006.

9. Kavraki, L.; Svestka, P.; Latombe, J.; Overmars, M. Probabilistic roadmaps for path planning in high-dimensional configuration spaces. *IEEE Trans. Robot. Autom.* **1996**, *12*, 566–580.

10. LaValle, S.M. *Rapidly-Exploring Random Trees: A New Tool for Path Planning.* Technical Report, TR 98-11, Computer Science Department, Iowa State University, USA, 1998.

11. Karaman, S.; Frazzoli, E. Sampling-based algorithms for optimal motion planning. *Int. J. Robot. Res.* **2011**, *30*, 846–894.

12. Islam, F.; Nasir, J.; Malik, U.; Ayaz, Y.; Hasan, O. RRT*-Smart: Rapid convergence implementation of RRT* towards optimal solution. In Proceedings of the International Conference on Mechatronics and Automation (ICMA), Chengdu, China, 5–8 August 2012.

13. Ferguson, D.; Likhachev, M. Efficiently using cost maps for planning complex maneuvers. In Proceedings of the ICRA Workshop on Planning with Cost Maps, Pasadena, California, USA, 19–23 May 2008.

14. Pivtoraiko, M.; Kelly, A. Fast and Feasible Deliberative Motion Planner for Dynamic Environments. In Proceedings of the ICRA Workshop on Safe Navigation in Open and Dynamic Environments: Application to Autonomous Vehicles, Kobe, Japan, 12–17 May 2009.

15. Koenig, S.; Likhachev, M. D* Lite. In Proceedings of the National Conference on Artificial Intelligence (AAAI), Edmonton, Alberta, Canada, 28 July–1 August 2002.

16. Lau, B.; Sprunk, C.; Burgard, W. Kinodynamic Motion Planning for Mobile Robots Using Splines. In Proceedings of the IEEE/RSJ International Conference on Intelligent Robots and Systems (IROS), St.Louis, MO, USA, 11–15 October 2009.

17. Walther, M.; Steinhaus, P.; Dillmann, R. Using B-Splines for Mobile Robot Path Representation and Motion Control. In Proceedings of the European Conference on Mobile Robots (ECMR), Ancona, Italy, 7–10 September 2005.

18. Sprunk, C.; Lau, B.; Burgard, W. Improved Non-linear Spline Fitting for Teaching Trajectories to Mobile Robots. In Proceedings of the IEEE International Conference on Robotics and Automation (ICRA), St. Paul, MN, USA, 14–18 May 2012.

19. Wilde, D.K. Computing clothoid segments for trajectory generation. In Proceedings of the IEEE/RSJ International Conference on Intelligent Robots and Systems (IROS), St. Louis, MO, USA, 11–15 October 2009.

20. Brezak, M.; Petrovic, I. Path Smoothing Using Clothoids for Differential Drive Mobile Robots. In Proceedings of the 18th IFAC World Congress, Milano, Italy, 28 August–2 September 2011.

21. Sprunk, C.; Lau, B.; Pfaff, P.; Burgard, W. Online Generation of Kinodynamic Trajectories for Non-Circular Omnidirectional Robots. In Proceedings of the IEEE International Conference on Robotics and Automation (ICRA), Shanghai, China, 9–13 May 2011.

22. Cirillo, M.; Uras, T.; Koenig, S.; Andreasson, H.; Pecora, F. Integrated Motion Planning and Coordination for Industrial Vehicles. In Proceedings of the 24th International Conference on Automated Planning and Scheduling, Portsmouth, NH, USA, 21–26 June 2014.

23. Pecora, F.; Cirillo, M.; Dimitrov, D. On Mission-Dependent Coordination of Multiple Vehicles under Spatial and Temporal Constraints. In Proceedings of the IEEE/RSJ International Conference on Intelligent Robots and Systems (IROS), Algarve, Portugal, 7–12 October 2012.

24. Cirillo, M.; Uras, T.; Koenig, S. A Lattice-Based Approach to Multi-Robot Motion Planning for Non-Holonomic Vehicles. In Proceedings of the IEEE/RSJ International Conference on Intelligent Robots and Systems (IROS), Chicago, IL, USA, 14–18 September 2014.

25. Pivtoraiko, M.; Knepper, R.A.; Kelly, A. Differentially Constrained Mobile Robot Motion Planning in State Lattices. *J. Field Robot.* **2009**, *26*, 308–333.

26. Pivtoraiko, M.; Kelly, A. Kinodynamic motion planning with state lattice motion primitives. In Proceedings of the IEEE/RSJ International Conference on Intelligent Robots and Systems (IROS), San Francisco, CA, USA, 25–30 September 2011

27. Likhachev, M.; Gordon, G.; Thrun, S. ARA*: Anytime A* with provable bounds on sub-optimality. *Adv. Neural Inf. Process. Syst.* **2003**, *16*, 767–774.

28. Knepper, R.A.; Kelly, A. High Performance State Lattice Planning Using Heuristic Look-Up Tables. In Proceedings of the IEEE/RSJ International Conference on Intelligent Robots and Systems (IROS), Beijing, China, 9–15 October 2006.

29. Houska, B.; Ferreau, H.; Diehl, M. ACADO Toolkit—An Open Source Framework for Automatic Control and Dynamic Optimization. *Optim. Control Appl. Methods* **2011**, *32*, 298–312.

30. Bock, H.; Plitt, K. A Multiple Shooting algorithm for direct solution of optimal control problems. In Proceedings of the 9th IFAC World Congress, Budapest, Hungary, 2–6 July 1984; pp. 242–247.

31. Munoz, V.F.; Ollero, A. Smooth trajectory planning method for mobile robots. In Proceedings of the Conference on Computational Engineering in Systems Applications, Lille, France, 9–12 July 1996.

4

Towards an Open Software Platform for Field Robots in Precision Agriculture

Kjeld Jensen [1,*]**, Morten Larsen** [2]**, Søren H. Nielsen** [1]**, Leon B. Larsen** [1]**, Kent S. Olsen** [1] **and Rasmus N. Jørgensen** [3]

[1] Faculty of Engineering, University of Southern Denmark, Campusvej 55, 5230 Odense M, Denmark; E-Mails: soeni05@gmail.com (S.H.N.); lelar09@student.sdu.dk (L.B.L.); keols09@student.sdu.dk (K.S.O.)

[2] Conpleks Innovation, Fælledvej 17, 7600 Struer, Denmark; E-Mail: morten.larsen@conpleks.com

[3] Institute of Engineering, Aarhus University, Nordre Ringgade 1, 8000 Aarhus, Denmark; E-Mail: rnj@iha.dk

* Author to whom correspondence should be addressed; E-Mail: kjen@mmmi.sdu.dk

Abstract: Robotics in precision agriculture has the potential to improve competitiveness and increase sustainability compared to current crop production methods and has become an increasingly active area of research. Tractor guidance systems for supervised navigation and implement control have reached the market, and prototypes of field robots performing precision agriculture tasks without human intervention also exist. But research in advanced cognitive perception and behaviour that is required to enable a more efficient, reliable and safe autonomy becomes increasingly demanding due to the growing software complexity. A lack of collaboration between research groups contributes to the problem. Scientific publications describe methods and results from the work, but little field robot software is released and documented for others to use. We hypothesize that a common open software platform tailored to field robots in precision agriculture will significantly decrease development time and resources required to perform experiments due to efficient reuse of existing work across projects and robot platforms. In this work we present the FroboMind software platform and evaluate the performance when applied to precision agriculture tasks.

Keywords: field robots; precision agriculture; FroboMind; ROS

1. Introduction

Robotics in precision agriculture has the potential to improve competitiveness and increase sustainability compared to current crop production methods [1], and has become an increasingly active area of research during the past decades. Tractor guidance using Global Navigation Satellite System (GNSS) based sensor systems for route following and local sensor systems for accurate in-row navigation in row crops and orchards have already reached the market. One example is the John Deere iTEC Pro which supports GNSS based steering in straight and curved rows and at headlands while controlling speed and performing active implement guidance. Another example is the Claas Cam Pilot system which navigates a tractor through row crops using 3D computer vision to detect the location of the crop rows. Early prototypes of smaller field robots performing precision agriculture tasks without human intervention also exist [2–6]. But research in advanced cognitive [7] perception and behaviour that are required to enable a more efficient, reliable and safe autonomy becomes increasingly demanding. The level of complexity in an unstructured, dynamic, open-ended and weather influenced environment like a crop field or an orchard is high, and the size and complexity of the software needed to perform experiments impose ever greater demands on research groups. A lack of collaboration between the research groups contributes to the problem. Scientific publications are published on findings and results in precision agriculture, but little field robot software has actually been released, published and documented for others to use. We hypothesize that a common open software platform tailored to field robots in precision agriculture will significantly decrease the development time and resources required to perform field experiments due to efficient reuse of existing work across projects and robot platforms. The aim of this work is to establish such a software platform and evaluate the performance when applied to different precision agriculture tasks and field robots.

Related Work

The literature contains numerous references to relevant proposals and implemented solutions within robot software architectures, frameworks, middlewares, development environments, libraries *etc*. Below is a brief review of some of the best known and widely used solutions. Table 1 compares the middleware specifications.

CARMEN Robot Navigation Toolkit from the Carnegie Mellon University (CMU) [8] is a modular software library for mobile robot control. It provides interfaces to a number of robot platforms and sensors, a 2d simulator and algorithms for localization, mapping, path planning *etc*. The architecture features 3 layers, the lowest layer contains hardware interfaces and collision detection, the middle layer localization and navigation, and the highest layer contains all high level tasks. Inter Process Communication (IPC), another project by CMU, is used for sending data between processes based on TCP/IP sockets.

CLARAty (Coupled Layer Architecture for Robotic Autonomy) [9,10] is a framework for generic and reusable software for heterogeneous robot platforms developed by the Jet Propulsion Laboratory. CLARAty is a two-tiered coupled layer (decision and functional) architecture. The functional layer is a modular software library which provides an interface to the robot system and contains algorithms for low- and mid-level autonomy. The decision layer builds on top of this adding high-level autonomy to achieve mission goals. CLARAty consists of a public and a private repository.

Table 1. Comparison of middleware specification. The year listed under *Update* indicates the latest official release or substantial update.

	CARMEN	**CLARAty**	**MRDS**	**ORCA**	**Orocos**	**Player**	**ROS**
Updated	2008	2007	2012	2009	2014	2010	2014
Main Languages	C	C++	C# VPL	C++	C++	C++ TCL Java Python	C++ Python Lisp
Primary Platforms	Linux	VxWorks Linux Solaris	Windows	Linux	Linux Windows OSX	Linux Solaris BSD OSX	Linux
Component Interface	IPC	Unknown	CCR, DSS	ICE	CORBA	TCP sockets	XMLRPC
License	GPLv2 /BSD	Proprietary /Closed	Commercial /Academic	LGPL/ GPL	LGPL	GPLv3	BSD 3-Clause

MRDS (Microsoft Robotics Developer Studio) [11] is a development environment for robot control and simulation. MRDS has support for a number of programming languages including it's own platform specific language Visual Programming Language (VPL). It supports a wide range of robotics hardware platforms and integrates a fully featured simulation environment. The component interface is based on the .NET Concurrency and Coordination Runtime (CCR) library for managing asynchronous, parallel tasks using message-passing and Decentralized Software Services (DSS), a lightweight .NET-based runtime environment.

Orca [12] is an open-source software framework for developing component-based robotic systems. The aim of Orca is to promote software reuse through definition of interfaces, providing component libraries and maintaining a public component repository. The intended use is for both commercial applications and research environments. The Orca project is a branch out from Orocos, one main difference is that Orca uses the Internet Communications Engine (ICE) [13].

Orocos (Open Robot Control Software) [14] contains portable C++ libraries for advanced machine and robot control. Orocos builds upon the open source Common Object Request Broker Architecture (CORBA) middleware. Orocos is in active development and contains extensions that support other frameworks including ROS.

Player [15] is one of the most used robot control interfaces and supports a wide variety of robots and components. The Player architecture is based on a *network server* which runs on the robot platform and provides a TCP socket interface to the robot. The *client program* is thus able to read data from sensors, write commands to actuators *etc.* by connecting to the socket. Player coexists with Stage which is a 2d multiple robot simulator and Gazebo which is 3d multiple robot simulator based on a physics engine.

ROS (Robot Operating System) [16] is a flexible framework for writing robot software. It provides services, libraries and tools for building robotics applications. Examples are hardware abstraction and device control, interprocess communication, multi-computer environment support *etc*. ROS is in active development and is maintained by Open Source Robotics Foundation who provides access to a large repository of available components and maintains a list of external repositories provided by the growing community. The ROS core code and most ROS packages are released under the BSD 3-Clause License which facilitates commercial use of the code.

In terms of robot software architectures the above mentioned *Carmen* and *CLARAty* each use their own architecture, wheres a *MRDS*, *ORCA*, *Orocos* and *ROS* don't specify or endorse a certain architecture. *Player* does not specify an architecture beyond the two-layer *Network server* and *Client program*. Several relevant but lesser known robot software architectures are described in the literature:

Ref [17] argues that the sense-model-plan-act paradigm used in most architectures is unable to react in a dynamical environment because it depends heavily on the model, and that a behaviour-based approach such as the subsumption architecture [18] is limited by the purely reactive behaviours and does not perform well when carrying out complex tasks. A hybrid is thus presented containing three layers: Hardware layer, Reactive layer and Deliberative layer. The reactive layer is grouped into a perception module (localization and mapping) and an action module (navigation and actuation). The deliberative layer contains high level mapping and path planning.

Ref [19] introduces Agricultural Architecture (**Agriture**), a control architecture designed for teams of agricultural robots. Agriture consists of three layers, the physical layer which is either the robot or the Stage/Gazebo simulators, an Architecture middleware based on *Player*, *Java Agent Development Framework* and *High Level Architecture*, and a Distributed application layer for the application software.

Ref [20] proposed **Agroamara**, a hybrid agent of behaviour architecture for autonomous farming vehicles. Here *agent of behaviour* describes computational and control processes addressed to reach or to maintain a goal, with perceptual, deliberative and reactive abilities. The architecture groups the agents into perceptual and motor agents and uses layers to describe the information flow and hierarchy of activation of the agents. Data sharing is handled through shared memory and peer to peer message parsing. Multiprocessing is supported using winsocket.

Ref [21] describes the software architecture of the Autonomous Mobile Outdoor Robot (**AMOR**) which won the first price in autonomous driving in the European Land Robot Trial (ELROB) 2007 for urban and non-urban terrain. The robot software is not publicly available, however the paper does provide a good introduction to the design as well as the navigation modules. Intercommunication is handled by a *Virtual Sensor Actor Layer* that use TCP sockets and unifies the access to all sensors and actors. The main modules of the high level software is *global model* (map database), *global path planning* (deciding plans for the indended behaviour), *local model* (based on laser scanner and camera data), *local path planning* and *basic reflexes*.

Ref [22] introduces **Mobotware**, a plug-in based framwork for mobile robots. Mobotware is divided into a hard- and a soft realtime section. Each section is decomposed into layers of increasing abstraction: Hardware abstraction, reactive execution, deliberative perception and planning. The framework has three core modules, a *Robot Hardware Daemon* providing hardware abstraction, a *Mobile Robot Controller* handling real-time closed loop controlling, and *Automation Robot Servers* handling sensor processing, mission planning and management.

Ref [23,24] propose a system architecture to enable behavioural control of an autonomous tractor based on the subsumption architecture [18]. This work is related to the Software Architecture for Agricultural Robots (**SAFAR**) project with the purpose to develop a set of designs, tools and resources to promote the development of agricultural robots. SAFAR is based upon MRDS and is closed source but supports a Python scripting engine.

In 2005 the Stanford robot **Stanley** won the DARPA Grand Challenge by driving autonomously for 131 miles along an unrehearsed desert trail [25]. The robot software is not publicly available, however the paper provides information on the software architecture and design considerations. Stanley contains approximately 30 software modules organized in 6 layers representing sensor interfaces, perception, planning and control, vehicle interface, user interface and global services. All modules are executed individually without interprocess synchronization and all data are globally time stamped. Interprocess communication is handled using publish-subscribe mechanisms based on the CMU IPC kit.

Table 2 compares some properties of the reviewed architectures with respect to this work. The data is based on the reviewed publications and information publicly available on the web. It should be noted that all the reviewed publications contains valuable and relevant knowledge, but at the same time the table underpins the need for collaboration between the research groups in precision agriculture with respect to software reuse as discussed in the problem description.

Table 2. Comparison of robot software architectures with respect to the problem domain and hypothesis of this work. In the columns Agricultural applications, Multiple platforms and Multiple users *Yes* means that practical field trials have taken place. (*Yes*) in Open source means that not all software has been released and/or the license is not permissive free.

	Agricultural Applications	Multiple Platforms	Multiple Users	Open Source	Updated Recently
CARMEN	Yes	Yes	Yes	(Yes)	No
CLARAty	No	Yes	Yes	(Yes)	No
Agriture	No	No	No	No	No
Agroamara	Yes	No	No	No	No
AMOR	No	No	No	No	No
Mobotware	Yes	Yes	Yes	No	Yes
SAFAR	Yes	Yes	No	No	No
Stanley	No	No	No	No	No

2. The FroboMind Software Platform

2.1. Design Goals

The software platform presented in this work has been named FroboMind which is a contraction of *Field Robot Mind*. It is based on design goals that promote reuse, support projects of varying size and complexity, and facilitate collaboration:

- **Modularity:** It must provide a proper modular structure to ease the development of new components and facilitate reuse. Interfaces between modules must be well defined, however still allowing researchers and developers to experiment freely.
- **Extensibility:** The design must support a wide variety of applications from a student working with signal processing or route planning algorithms to large scale field experiments conducted by research groups with several robots interacting online with each other and human operators.
- **Scalability:** It must support online computation of high load algorithms such as computer vision and mapping with respect to both memory and processing power requirements. This includes scaling to distributed applications on networked heterogeneous platforms.
- **Software reuse:** There must be a strong focus on minimizing resource requirements in the development of new applications by facilitating reuse of software. The future workload of maintaining the software platform is also an important consideration. When establishing the software platform, existing work should be reused to the extent possible. To prevent maintenance problems arising due to the existing work, this must be under active development and supported by a community.
- **Open source:** All core components of the software platform must be released under a permissive free software license to keep it free for others to use, change and commercialize upon. This is necessary to facilitate collaboration with industrial partners.

2.2. FroboMind

The FroboMind software platform structure consists of four parts as illustrated in Figure 1. The *Operating System* provides basic facilities for application execution, file handling, hardware interfacing, communication, networking *etc*. The *Middleware* provides services like timing and communication between software components in a distributed system which simplifies the software development significantly. The *Architecture* organizes the software components into layers and modules with well defined interfaces. The modularity eases the development proces and facilitate efficient software reuse. The *Components* are the actual building blocks used for field robot applications and reused across projects and robot platforms. In the following sections each of the four parts of the software platform structure is described in detail.

Figure 1. Overview of the FroboMind software platform structure.

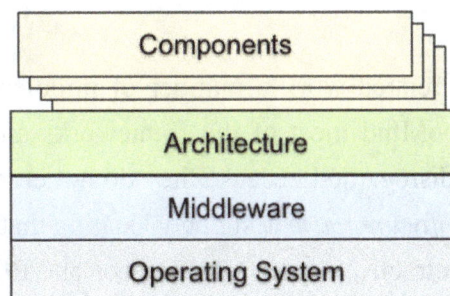

2.3. Operating System

When choosing an operating system for FroboMind the design goals, in particular the *Software reuse* and *Open source* goals, must be taken into consideration.

In robotics software implementations a Real-Time Operating System (RTOS) dedicated to embedded systems is typically used. The reason is that RTOS improves the performance for components that have hard or near real-time requirements, and some of the available systems (e.g., VxWorks and QNX) improve the reliability significantly compared to traditional operating systems. However they are usually not open source, and software developed for the RTOS typically runs on the target system only and therefore requires cross compilation, and on-target debugging *etc.* Linux based open source RTOS exist in different varieties. Examples are dedicated distributions where the entire system run as preemptive processes (e.g., RTLinux), add-ons that create a small microkernel where linux runs as a task, and kernel patches or libraries that implement a near real-time environment like the RealTime Application Interface for Linux (RTAI) [26] and CONFIG_PREEMPT_RT [27]. Some of these are well functioning and are utilized in commercial products as well as academic projects. However they are trying to match an RTOS with a fully fledged operating system which is inherently a tradeoff between low latency requirements and the overall efficiency of the system [28]. Other typical challenges of the RTOS are small development teams and a limited community supporting the projects. The RTOS distributions therefore often lag behind regarding support for new software and hardware, and support options in case of problems are limited. In applications where the robot is connected to the internet, network security may be an issue as well due to delayed software updates.

Based on the above considerations the linux distribution Ubuntu was chosen as operating system for FroboMind. Ubuntu is one of the largest and most active linux distributions available, and the annual Long-Term Support versions have guaranteed support for five years after release. Choosing Ubuntu has the major advantage that one can effortlessly run the same operating system on the field robot and a standard laptop used for software development. There is also a huge community of linux software developers which is an advantage when support is needed for a particular problem. Ubuntu only supports soft real-time execution, but in field robotics real-time requirements at the order of 50–200 Hz typically only exist in relation to low level actuator controlling, and this task is often distributed to an external embedded controller that communicates via a serial or network interface. The use of Ubuntu in robot systems with near real-time requirements is evaluated in the experimental section.

2.4. Middleware

As described in the review of related work a number of middlewares designed for robot control exists already. However for FroboMind most of the frameworks and middlewares listed in Table 1 and described elsewhere must be disregarded because they do not comply with the *open source* design goal. Considering in addition the *software reuse* design goal stating that the middleware must be in active development this leaves only two attractive contestants: Orocos and ROS.

Taking a closer look at Orocos it is based upon a few principal components: The Orocos Toolchain which contains the tools required to build Orocos components, the iTaSC framework (generates robot motions using constraints). the rFSM toolkit (Finite State Machines) and the libraries KDL (Kinematics and Dynamics) and BFL (Bayesian Filtering). The Orocos Toolchain includes the Real Time Toolkit (RTT) framework for developing real-time components in C++.

The ROS middleware offers inter-process communication through a set of facilities: Anonymous publish/subscribe message passing, recording and playback of messages, request/response remote procedure calls and a distributed parameter system. ROS also provides a set of robot libraries such as robot geometry, pose estimation, localization, mapping and navigation *etc*. ROS has a very good support for distributing components between networked computers. ROS is not a realtime framework and is thus not suitable for hard real-time critical software such low level control algorithms. There is, however, support for integration with the Orocos RTT.

Orocos and ROS both fit well with the FroboMind design goals. In favor of Orocos it supports real-time execution and the code base appears to be stable. The requirements for CPU power and memory appear to be lower than for ROS. But Orocos seems more oriented towards low level control than robot application building, and it is not as user friendly as ROS, which also has by far the largest supporting community and the most active development. ROS provides a very detailed documentation using a wiki, videos and examples. Fast prototyping using scripting is available with Python which may speed up application development in many cases. Based on these considerations ROS was chosen as middleware for FroboMind. The requirements with respect to computing power when running ROS and Ubuntu in robot systems with near real-time requirements is evaluated in the experimental section.

2.5. Architecture

The purpose of having a unified architecture shared across different projects and research groups is to facilitate software reuse at the application level. This allows the entire software platform to be transferred to new applications and field robots, the user needs only to add new high level behaviours and low level interface drivers needed for a particular application.

ROS does not propose or endorse a certain architecture or structure of the software, instead the different software components named ROS nodes are able to intercommunicate freely without restrictions to structure. As a result different users tend to use different designs and to some extent different interfaces between components. The user therefore often has to build a new robotic application from scratch, and reuse mainly exists from component level down to snippets of code.

Adding a unified architecture to FroboMind increases the reusability of the components, but at the same time it limits the flexibility that ROS provides. In this section an architecture is proposed with the aim of balancing the tradeoff between reusability and flexibility.

Designing a conceptual architecture that appears intuitive to researchers, engineers and technicians is a challenging task due to their different backgrounds and experience from different projects. A simple basis for mobile robot software has been formulated by the questions: *Where am I?*, *Where am I going?* and *How should I get there?* [29]. The robot localizes itself based on percepts and prior knowledge, it then plans a route which leads towards the mission goal and navigates the route using motion control (Figure 2). However for a field robot in precision agriculture the task is the primary objective and navigation is merely a means to fulfil the task. Performing the task typically involves interaction with an attached or trailed implement, and the conceptual architecture in Figure 2 is therefore insufficient.

Instead a more generic approach based on Intelligent Agents [30] is used for modelling the architecture. An agent is an autonomous entity which perceives its environment through sensors and acts upon that environment through actuators. The action taken by the agent in response to any percept sequence is defined by an agent function. The FroboMind principal architecture consists of perception, decision making and action layers (Figure 3a). The layers have been decomposed to define abstraction levels (Figure 3b).

Figure 2. A conceptual basis for the architecture of mobile robot navigation software described in the literature: *Where am I? Where am I going? How do I get there?*

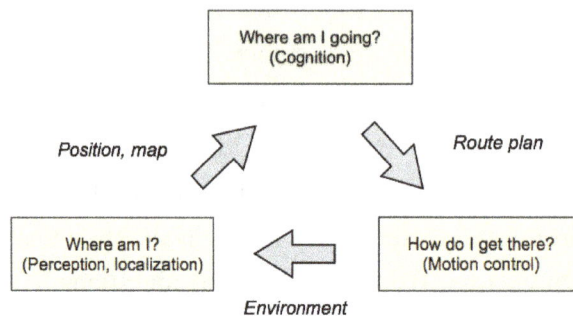

Figure 3. Decomposition of an Artificial Intelligence agent. (**a**) The agent perceives its environment through sensors and acts through actuators. The action taken by the agent in response to any percept sequence is defined by an agent function. (**b**) Decomposition to define data- and hardware abstraction levels.

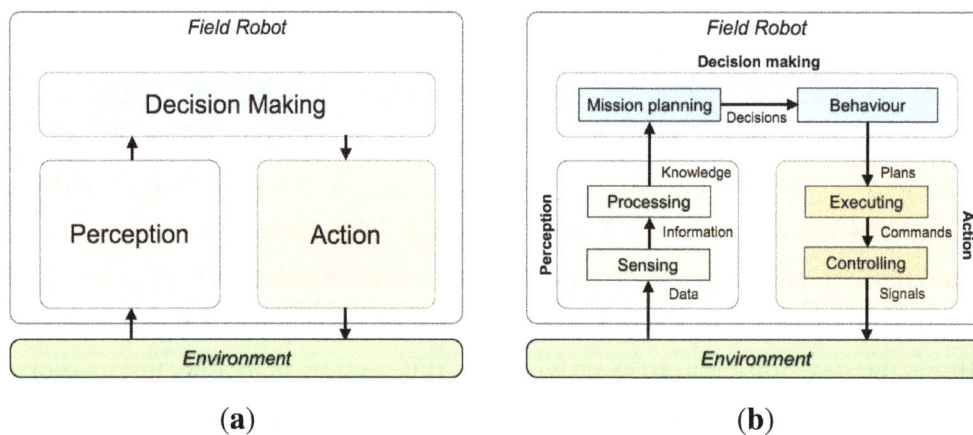

(**a**)   (**b**)

Perception represents the organization, identification and interpretation of sensory information in order to represent and understand the environment [31]. The perception layer has been decomposed into:

- **Sensing:** The robot perceives the surrounding partially observable environment (external percepts) through its sensors and assess the system interior state (internal percepts) through feedback from the platform and implement systems. Together these percepts constitute the available *information*. In addition this layer constitutes the abstraction between sensor hardware and the other parts of the architecture.

- **Processing:** By combining this *information* with previous, shared and a priori knowledge, the field robot maintains a model of the world and system state which constitute the accumulated *knowledge*. Since observables are governed by a certain amount of uncertainty, probabilistic methods are typically utilized to optimize the model.

Decision Making constitutes the cognitive layer and represents an AI agent function which determines the action taken in response to the accumulated *knowledge*. The implementation of the decision making layer may vary depending on application and users may choose to use other methodologies than the current support for model-based, utility-based AI agents and hierarchically organized finite-state machines to describe robot behaviour.

- **Mission Planning:** The robot mission planner continuously monitors the accumulated knowledge as well as any user interaction, and makes on this basis a *decision* about the optimal behaviour which leads towards the fulfillment of the mission. This corresponds to an AI agent utility function described in [30].
- **Behaviour:** The optimal behaviour decided by the mission planner origins from a library of possible behaviours available to the robot and implement. The active behaviour continuously monitors the accumulated knowledge and user interaction and updates action *plans* for the robot and implement action accordingly.

Action The action layer carries out the action defined by the action plans and has been decomposed into:

- **Executing:** The *plans* produced by the active behaviour are executed with respect to time and state. Based on this *commands* are sent to the controlling layer.
- **Controlling:** Commands issued by the executing layer are transmitted to low level controllers within the architecture or to external controller interfaces. This layer constitutes the abstraction between the field robot actuator hardware and other parts of the architecture.

The FroboMind architecture (Figure 4) represents an expansion of the decomposition in Figure 3. This is indicated by the grouping of modules containing software components into the *Perception*, *Decision making*, *Action* and *Safety* layers as well as the data interface between layers and modules indicated by the blue dashed lines. Internal fault diagnosis and incident handling are organized in a separate *Safety* module to minimize potential software errors and thus ensuring a high level of reliability. In order not to clutter the overview it is assumed that any component has access to data accessible by its predecessor. Multiple connections to successors are shown only where relevant to the understanding, and data available globally has not been included.

2.6. Components

The FroboMind components are implemented as ROS packages. Most of the current lower level components at the Perception and Action layers are written in C++ while many of the higher layer components are written in Python.

The components are located in a directory structure where each layer in the architecture (Figure 4) is represented by a directory (Table 3), each module as a subdirectory herein and the components are located in the module subdirectories. The FroboMind component directory structure and related documentation are located in a repository available through the FroboMind website [32].

Figure 4. The FroboMind Architecture. The modules containing software components are grouped into the *Perception*, *Decision making*, *Action* and *Safety* layers. The blue dashed lines indicate the data interfaces between layers and modules.

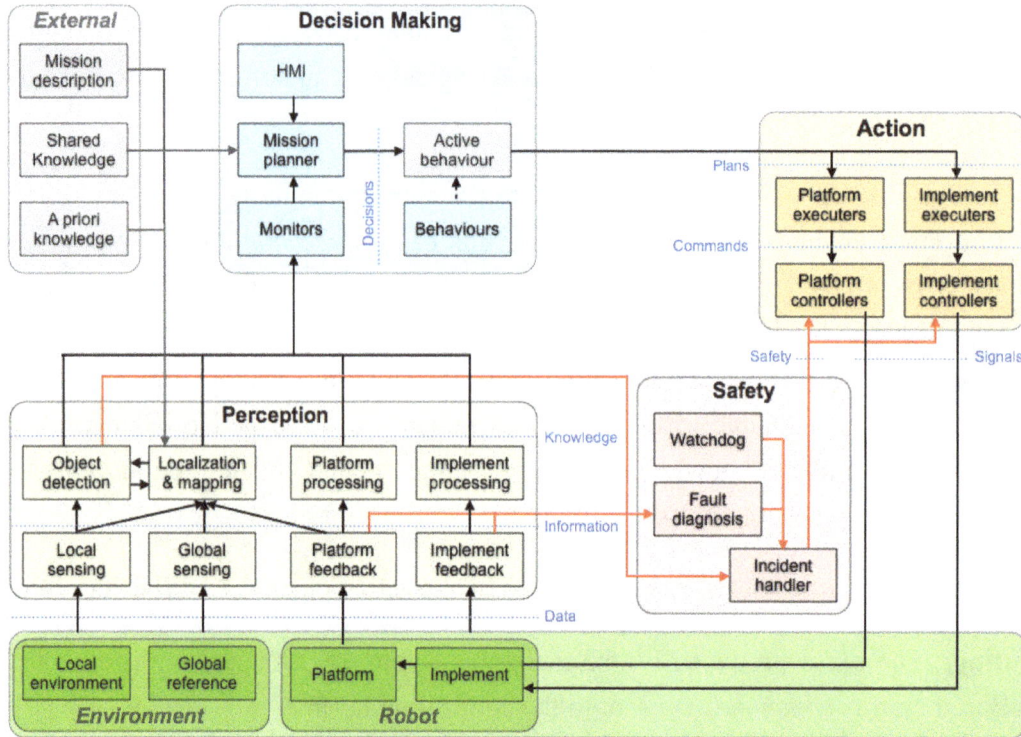

Table 3. The component (top level) directory structure at the FroboMind repository.

Directory	Content
/fmSensors	Sensor interfaces and information extraction (hardware abstraction layer)
/fmProcessors	Processing sensor information to obtain knowledge about the robot state.
/fmDecisionMakers	Robot mission, behaviour and HMI components.
/fmExecutors	Executing the current behaviour, e.g., navigation and implement control.
/fmControllers	Low level controllers and actuator interfaces (hardware abstraction layer).
/fmSafety	Watchdog, fault diagnosis and incident handling.
/fmApp	Directories containing components, scripts, launch files *etc.* related to the application of FroboMind to a particular project.
/fmLib	Interface drivers, software libraries *etc.* not represented in the architecture.
/fmTools	Various non-ROS tools and utilities.

2.7. Versions

FroboMind has been in active development since 2011 where the first prototype architecture and components were implemented in ROS Electric on Ubuntu 10.04. Since then FroboMind versions have been created each year with an updated architecture, improved components and compatible with updated versions of Ubuntu and ROS. The version numbers corresponds to the farming season. Version 2014 is the current FroboMind software platform described in this work. It is based on ROS Hydro and Ubuntu 12.04 LTS.

3. Experimental Section

In this work three experiments have been carried out to evaluate the performance of FroboMind when applied to various precision agriculture tasks and field robots. The first experiment evaluates the performance of FroboMind in robot systems with near real-time requirements including requirements with respect to computing power. The second experiment evaluates whether FroboMind significantly decreases the development time and resources required to perform field experiments. The third experiment uses available data from existing projects and robotic platforms that use FroboMind, to evaluate software reuse across these projects.

3.1. Evaluating the Performance of FroboMind in Robot Systems with Near Real-Time Requirements

An experiment was defined to evaluate the performance of FroboMind including ROS and Ubuntu in robot systems with near real-time requirements including requirements with respect to computing power. The experiment is based on a typical application for a field robot in precision agriculture: Autonomous navigation of a predefined route while controlling an attached or trailed implement. Figure 5 shows an example of the FroboMind architecture listing the components required for the autonomous navigation. Each of the components are described in Table 4. All the library functions listed in Figure 5 are implemented in C++.

Figure 5. FroboMind example architecture for autonomous navigation of a route plan.

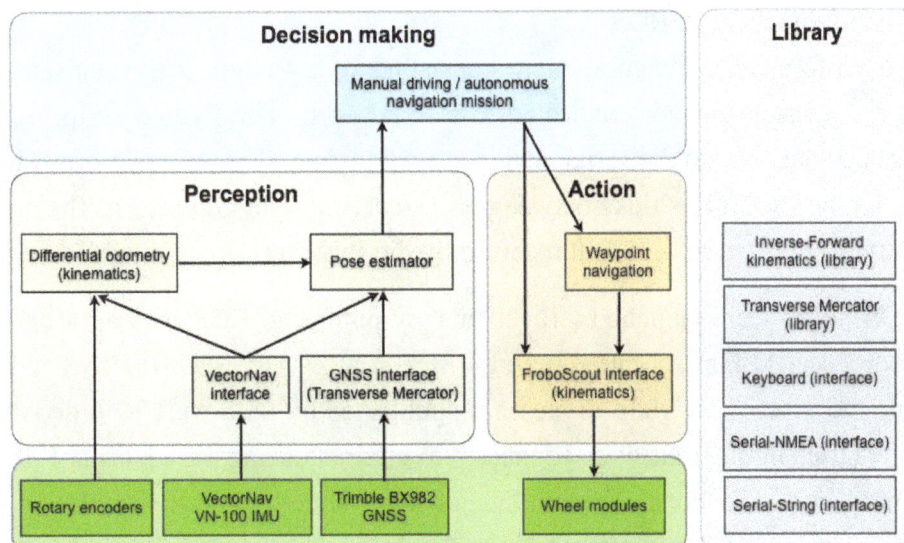

Table 4. FroboMind example components used for autonomous navigation of a route plan. Rate describes the update function rate. Source Lines Of Code (SLOC) include comments and blank lines but not library functions. c refer to C++ and p to Python.

Layer	Component	Rate	SLOC
/fmSensors	**VectorNav interface** reads and publishes the yaw axis gyro rate from a VectorNav VN-100 IMU.	40 Hz	261c
	GNSS interface reads and publishes the GNSS *GPGGA* information including conversion of the position to Transverse Mercator coordinates.	10 Hz	271c
/fmProcessors	**Differential odometry** estimates the current position and orientation based on information from the wheel encoders and the IMU yaw angle gyro.	50 Hz	527c
	Pose estimator uses pre-processing and an Extended Kalman Filter to estimate the absolute position and orientation based on the differential odometry and GNSS information.	50 Hz	592p
/fmDecisionMakers	**Manual driving/autonomous navigation mission** In the *manual driving* behaviour velocity commands from the keyboard are sent directly to the FroboScout interface. In the *autonomous navigation* behaviour the waypoint navigation component is activated.	20 Hz	199p
/fmExecutors	**Waypoint navigation** navigates the waypoint list by publishing linear and angular velocity commands.	20 Hz	1077p
/fmControllers	**FroboScout interface** converts linear and angular velocity commands to wheel velocities, publishes encoder feedback.	50 Hz	428p

In the experiment a differentially steered FroboScout robot (Table 5) was set to autonomously navigate a static route plan containing 5 waypoints using the architecture in Figure 5. The route length is 47 m. Two auxiliary components were added to monitor the system real time performance:

- A computer load monitor component which averages the CPU and memory load every 0.2 s and publishes the results through ROS.
- A real time performance test component consisting of a Python script and an external Atmel ATmega series microcontroller connected to a serial port. The Python script running at at rate of 100 Hz sends one byte to the serial device at each update. The ATmega firmware continuously counts the number of 100 μs ticks between each received byte and returns the latest count upon receiving a byte. This count is in turn received by the Python script and published through ROS.

In total 14 ROS nodes were launched. Together they published ROS messages under 28 different ROS topics, hereof 2 at 100 Hz, 9 at 50 Hz, 1 at 40 Hz, 4 at 20 Hz, 3 at 10 Hz, 6 at 5 Hz and 3 at a lower rate. During the experiment trials all messages published by ROS were recorded using the rosbag tool. The computers used in the trials were connected to internet during the trials but FroboMind did not initiate external network connections. All computers had a standard Ubuntu installation, no attempts to optimize for performance by shutting down default services were made.

Table 5. A list of robots using FroboMind. Column 3 lists the FroboMind version number followed by the reused components detailed in Table 6.

	Latest Application	**FroboMind**
ASuBot 1	Navigating orchards using lidar.	2013 ASU RDT RNV
Armadillo I 2	Navigating maize fields using lidar.	2012 ARD RDT RNV
Armadillo Scout 3	Crop monitoring.	2012 ARD
Armadillo III 4	Driving and navigation in dunes.	2012 ARD ODO POS WPT
AMS 5 Figure 6a	Navigating orchards using lidar [33].	2012 AMS RDT RNV
Armadillo IV 6	Large scale precision spraying trails [34].	2013 ARD ODO POS WPT PMP
Pichi 7	Detection and mapping of anti-tank mines and large unexploded objects.	2013 PIC ODO POS WPT COV MDI HMP
Robotti 8	Mechanical row weeding [2].	2013 ARD ODO POS WPT RWD
FroboMower 9	Mowing using low cost sensors.	2013 FBT ODO POS WPT
Frobit 10	Education [35] & rapid prototyping.	2013 FBT ODO WPT
SMR 11	Education.	2013 MBW ODO WPT
GrassBots 12	Grass mowing on low lands.	2013 GBT ODO POS WPT COV
FroboScout 13	Highway surveying.	2014 FST ODO POS WPT SUR

Table 6. List of components referred from Table 5. Physical Source Lines Of Code (SLOC) for the Grassbot robot interface is unknown as it is closed source. c refer to C++ and p to Python.

Component	**Description**	**SLOC**
AMS	Interface to the AMS robot	410c
ARD	Interface to the Armadillo robot	845c
ASU	Interface to the ASuBot robot	435c
COV	Area coverage algorithm	170p
FBT	Interface to the Frobit robot	285c
FST	Interface to the FroboScout robot	428p
GBT	Interface to the Grassbot robot	?
HMP	Hazard mapping algorithm	850c
MBW	Interface to Mobotware [22]	367c
MDI	Mine Detection Implement interface	125c
ODO	Differential odometry algorithm	527p
PIC	Interface to the Pichi robot	890c
PMP	Parcel map localization algorithm	330p
POS	Pose estimator algorithm	592p
RDT	Detecting crop rows using LIDAR	266c
RNV	Navigation in row crops	255c
RWD	Row weeding implement interface	90c
SUR	Highway surveying application	304p
WPT	Waypoint navigation algorithm	775p

Figure 6. Autonomous Mechanisation System (AMS) and orchard used for the test. (**a**) AMS; (**b**) Apple three orchard.

(**a**)   (**b**)

3.2. Evaluating if FroboMind Significantly Decreases the Resources Required to Perform Field Experiments

Evaluating whether the FroboMind software platform significantly decreases the development time and resources required to perform field experiments is difficult because no reference data seem to exist in the literature. An experiment was therefore conducted with the purpose of trying to quantify the portability of FroboMind to a new robot platform by analysing the associated work. The experiment was conducted in collaboration with the University of Hohenheim (UH) and is presented in [33]. A brief description and the results is included here for completeness.

The experiment was designed so that as many parameters as possible were controllable. The task was defined as field robot navigation through an apple tree orchard using local sensors only which is another typical precision agriculture application [36,37]. The Autonomous Mechanisation System (AMS) robot (Figure 6a) owned by UH was used for the experiment. The AMS has been utilized in several previous precision agriculture research projects [38,39] using Mobotware [22], and both the hardware and low level software is well tested and documented. A lidar and an Inertial Measurement Unit (IMU) were used as navigation sensors. An RTK-GNSS system was included for the purpose of recording reference data.

The experiment was conducted during a 5 day workshop where 4 developers worked full time porting FroboMind to the AMS, implementing the orchard navigation task and documenting the process. Preparations before the workshop included preparing the AMS and related hardware, reading documentation and planning the workshop. The task at the workshop was therefore defined as interfacing FroboMind to the AMS and building an application supporting autonomous navigation along the tree rows of the orchard and perform headland turns at the end.

In the orchard used for the navigation tests the tree rows were approx. 100 m long and interspaced by 4 m (Figure 6b). In the first trial the AMS mission was to navigate autonomously between two rows of trees. At the end of the row the AMS would make a 90 degree left turn, drive straight, and make a 90 degree left turn which would position the robot in an adjacent row. It was decided to repeat this process continuously 12 times corresponding to 6 full rounds driving the same track. In the second trial the AMS mission was to navigate autonomously through the same orchard alternating between left and right turns. Estimation of position and orientation (pose) relative to the tree rows is handled by FroboMind components from a previous project: A Ransac based algorithm uses the lidar data to detect the tree rows and outputs the angle and offset relative to the rows. The angle is fused with the IMU yaw angle data using an Extended Kalman Filter, and the resulting pose is used to control the AMS steering wheels using a PID controller. When the lidar detects the ends of the tree rows, it switches to a turn state and performs the turn based on the IMU data. When the lidar detects the ajacent row, it switches back to in-row navigation.

3.3. Evaluating Software Reuse Across Existing Projects Using FroboMind

During the past three years FroboMind has been used in various projects in precision agriculture as well as in some projects in similar domains. Examples of tasks are navigation of route plans using GNSS, navigation in row crops and orchards using local sensors, control of passive and active implements and interfacing to autonomous implements. Examples of sensor interfaces implemented in FroboMind are encoders, GNSS, IMU, lidar, 3d lidar, 3d vision row camera, localization using stereo vision, total station and metal detector. Examples of actuator interfaces controlled by FroboMind are relays, brushed and brushless motors, servos, linear actuators, hydraulics propulsion and diesel engines. Together these tasks and interfaces support many of the operations required in precision agriculture and the potential for efficient software reuse is thus high.

The third experiment is based on available data from existing projects and field robots that use FroboMind. The purpose is to evaluate the extent of software reuse by looking at the approximate number of physical Source Lines Of Code (SLOC) shared across these projects. As the projects have been carried out during a period where the FroboMind architecture and core components were in very active development it is difficult to perform an accurate comparison. But it provides an indication of the degree of reusability.

4. Results

4.1. Evaluating the Performance of FroboMind in Robot Systems with Near Real-Time Requirements

Three different trials were conducted, each using one of the computers listed in Table 7. The Graphical User Interface (GUI) *Ubuntu desktop* was running on PC2 and PC3, but on PC1 it was shut down prior to the trial because of high CPU load.

In all trials did the robot complete the waypoint navigation task successfully in about 100 s. Figure 7 shows the CPU and memory load averaged at 0.2 s intervals. Figure 8 shows the scheduling delays experienced by the 100 Hz FroboMind component during each of the 3 trials, Table 8 shows statistical data for the delays. Due to lack of an absolute, accurate time source in the experiment setup the delay measurements have been calibrated for offset and skew errors based on statistical calculations and, the data may therefore slightly inaccurate. The skew calibration constants were verified using external frequency measurement.

Table 7. Specifications for the computers used for the 3 trials in the FroboMind real-time performance experiment.

	PC1	PC2	PC3
Laptop:	Acer Aspire One ZG5	IBM ThinkPad X61s	Acer Travelmate B113
RAM:	1 Gb	2 Gb	4 Gb
CPU (Intel):	Atom N270	Core 2 Duo L7300	Core i3-2377M
Clock:	1.60 GHz	1.40 GHz	1.50 GHz
Cores/siblings:	1/2	2/2	2/4
Cache:	512 kb	4096 kb	3072 kb

Figure 7. CPU & memory load during autonomous navigation of a route plan for each of the 3 trials. The red graphs shows % of max CPU load. The black graphs show usage % of the total memory.

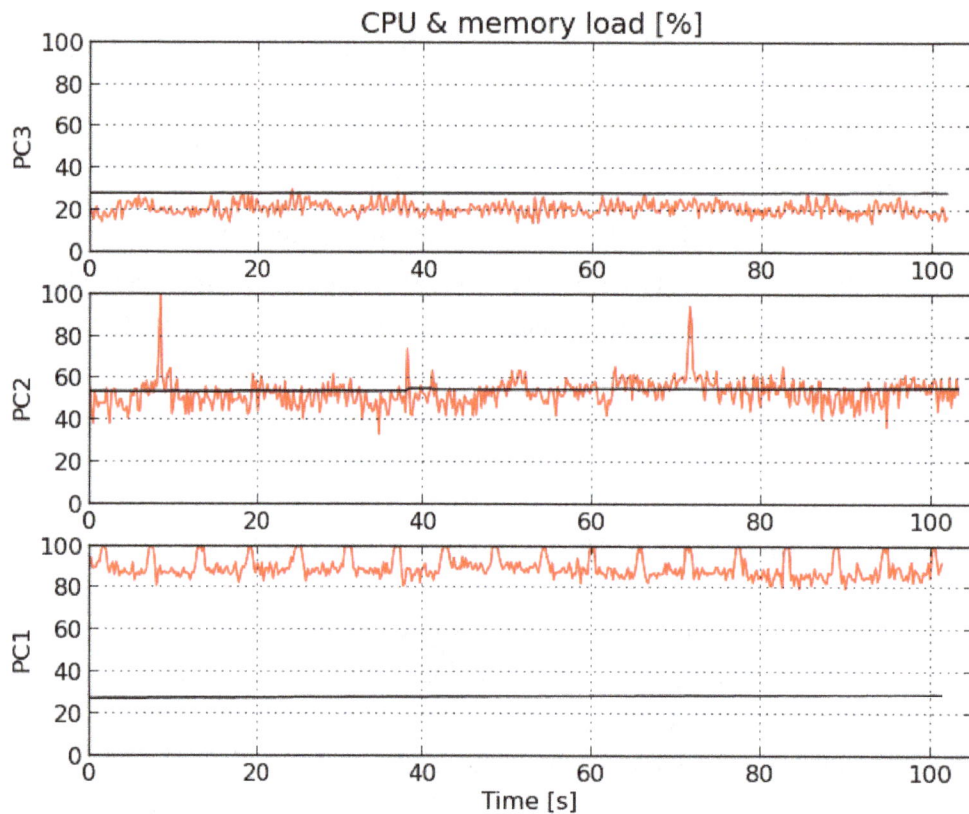

Figure 8. 100 Hz scheduler delays during autonomous navigation of a route plan for each of the 3 trials. The y axis represents the delay of each 10 ms schedule.

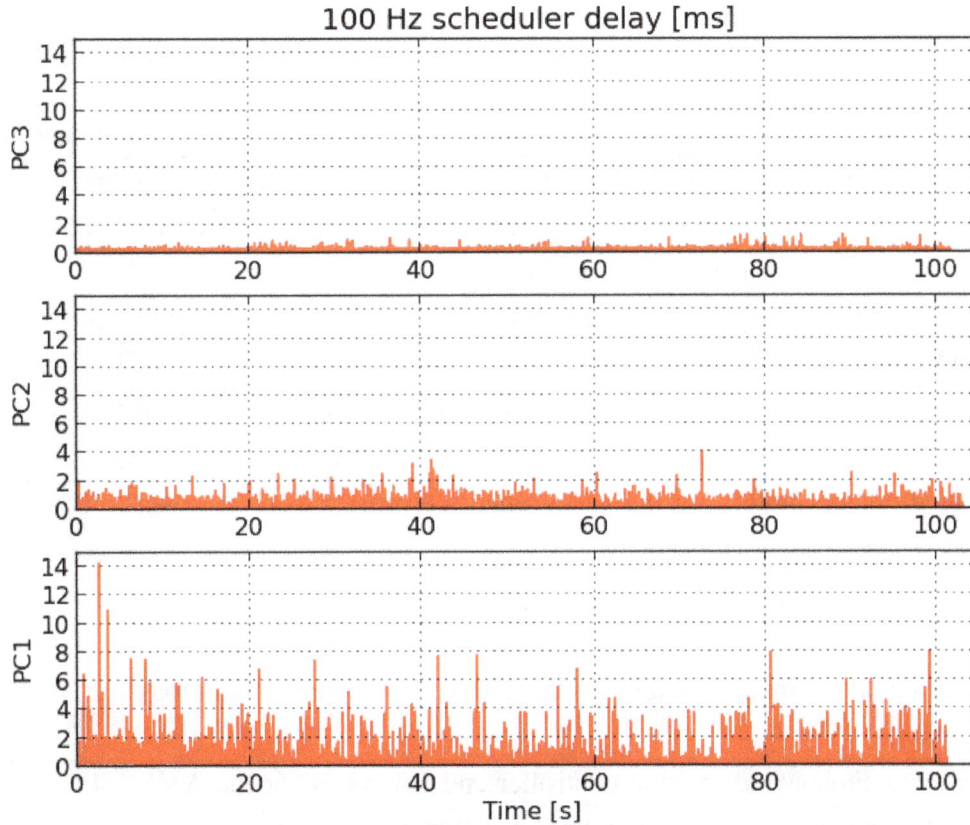

Table 8. Statistics for the schedule delays experienced by the 100 Hz component in the three trials of the FroboMind performance experiment.

	Mean	Variance	95'th Percentile	Maximum
PC1	0.58 ms	0.41 ms^2	1.52 ms	14.1 ms
PC2	0.26 ms	0.055 ms^2	0.72 ms	4.0 ms
PC3	0.24 ms	0.0074 ms^2	0.36 ms	1.3 ms

4.2. Evaluating if FroboMind Significantly Decreases the Resources Required to Perform Field Experiments

During the workshop a journal was continually updated with information about completed tasks *etc*. A summary of the estimated time consumption on different sub tasks is listed in Table 9.

While porting FroboMind to the AMS it was discovered that the AMS exhibited errors when controlling the front wheels. When requesting the AMS to drive straight it would drift slowly to the left. When requesting the AMS to turn left the wheels were positioned correctly but when requesting the AMS to turn right the steering wheels were positioned at an angle significantly lower than the requested angle. Based on the performed tests it is likely that the problem is with the steering hardware, however due to the time constraints it was decided to mitigate the problems by modifying the relevant FroboMind components. The process to identify, understand and mitigate these errors added substantially to the consumed time.

Table 9. Summary of estimated time consumption during the workshop.

Activity	Duration
Adding new components to FroboMind that directly support the AMS platform. This includes obtaining information about the AMS interface from documentation and available source code.	60 h
Simulating and testing remote controlled and autonomous driving with the AMS positioned in a test stand.	30 h
Adding and improving the documentation www.frobomind.org when the project work revealed issues not properly documented.	10 h
Updated documentation for the AMS when the project work revealed issues not properly documented.	10 h
Testing the implemented behaviours with the AMS on the ground.	20 h
Collecting data from a manual run in the orchard.	6 h
Analysing data from a manual run in the orchard to prepare autonomous operation.	20 h
Creating a FroboMind mission controller and behaviors for the AMS navigating autonomously in an orchard using local sensors.	40 h
Testing autonomous behaviour in the orchard.	30 h
Updating project documentation and web pages.	12 h

The first trial navigating 6 full rounds driving the same track in two adjacent rows was completed successfully. Three rounds were completed without human intervention. During two of the rounds the navigation failed one time at the same location where the apple trees at one of the rows were replanted by much younger and hence smaller trees. During one round the navigation failed at a location where trees were missing for more than 6 meters at one of the rows. Figure 9a shows the overlay GNSS track from all 6 rounds.

In the second trial navigating a section of rows in the orchard it was decided that focus would not be on the sensor/controller performance under difficult circumstances, so a person stepped in as "tree" whenever one of the rows had a large hole between the trees. The trial was concluded after completing 7 rows. The trial was completed without human intervention except two times where the AMS failed a turn to the right at a row end due to the described steering problems. Figure 9b shows the recorded GNSS track from this trial.

Figure 9. Recorded GNSS track from the trials. (**a**) shows the overlay GNSS track from all 6 rounds. (**a**) Trial 1; (**b**) Trial 2.

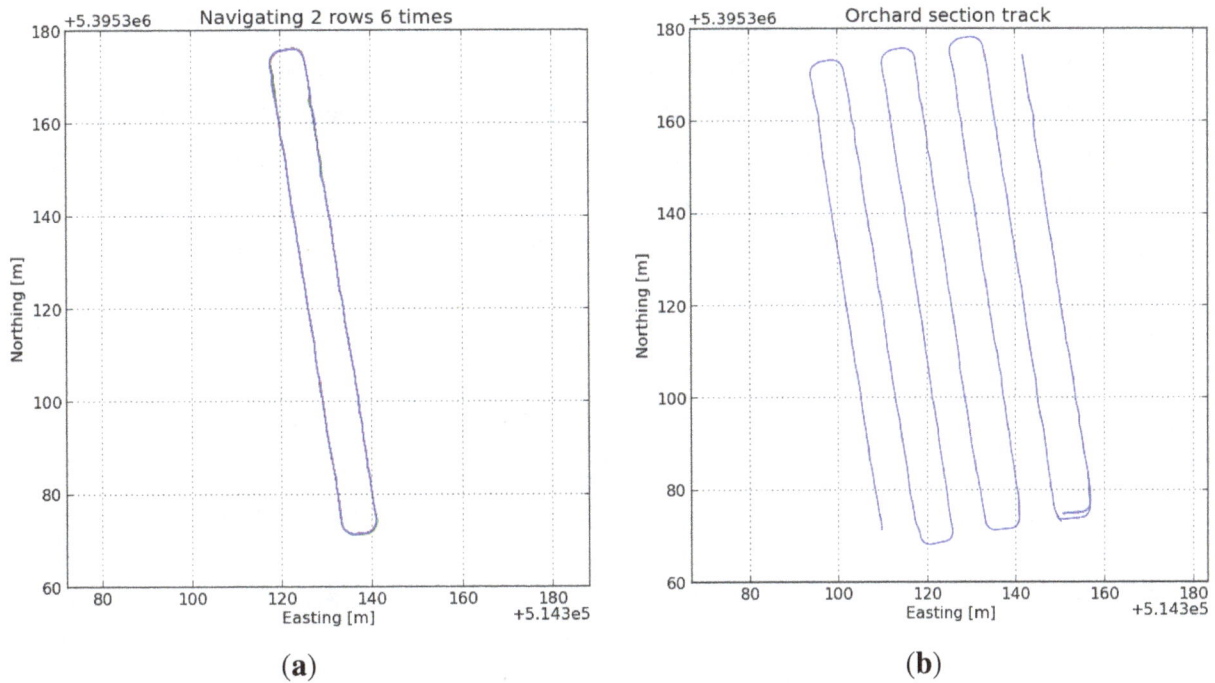

(**a**)

(**b**)

Figure 10. Images of robots using FroboMind (excluding the AMS depicted in Figure 6a). The numbers show the historical order of FroboMind integration and refer to the list of robots in Table 5.

4.3. Evaluating Software Reuse Across Existing Projects Using FroboMind

Table 5 lists robots that are known to have been interfaced to FroboMind. The table lists the latest known application where FroboMind has been used as well as the latest version of FroboMind known to work on the robot. Based on information obtained from each application, abbreviations for software components shared with other robots are listed next to the version number. The abbreviations refer to Table 6 which lists the component function along with the physical SLOC including comments and blank lines. Due to different component versions the SLOC may differ between different robot installations and should thus be considered to be approximate. Excluded from the list of components are sensor interfaces and libraries *etc*. Figure 10 shows images of each robot, the image number refer to the list of robots in Table 5.

5. Discussion

5.1. Evaluating the Performance of FroboMind in Robot Systems with Near Real-Time Requirements

Ubuntu was chosen as operating system and ROS as middleware because they complied best with the design goals *extensibility*, *scalability*, *software reuse* and *open source*. This was, however, at the cost of hard real-time capabilities that could have been obtained using e.g., RTAI [26] and Orocos [14]. Although hard real-time capabilities are rarely required in field robotics except for low-level control which is typically handled by external embedded systems, it is important to know the system timing uncertainty.

The computing power in the three computers used in the performance evaluation in the first experiment (Table 7) are at the low end of what is available today, however still above the currently popular ARM based embedded boards such as Raspberry Pi and BeagleBone Black.

The trials show a stable memory usage (Figure 7) which is consistent with that none of the active FroboMind components perform any significant dynamic memory allocation. The memory usage was approximately 1/3 Gb for PC1 where the GUI was disabled. For PC2 and PC3 the memory use was approximately 1 Gb.

The CPU load varies in the different trials which is to be expected considering the available computing power. The CPU load for PC1 exhibits a certain periodic pattern of peaks. A subsequent analysis revealed that the peaks were caused by an installed fan control script which sleeps 5 seconds between each execution, so this is not related to FroboMind.

Comparing the CPU load with the scheduler delays shown in Figure 8 and Table 8 it is clear that PC1 is pushed to the limits at this load and has difficulties running a 100 Hz scheduler without overstepping every now and then. PC2 and PC3 manages fine at the current load if the observed delays are acceptable to the application.

In other precision agriculture applications there may be components causing the CPU load to vary significantly with time which may be more or less predictable. Examples are dynamical mapping, route planning, image processing *etc*. In these cases there are different options: (1) Perform a similar experiment to validate that enough computing power is available to the application; (2) Distribute the components with varying load to another networked platform; (3) Integrate ROS with the Orocos Real Time Toolkit which is documented at the ROS website.

The results from this experiment gives rise to the recommendation that in any application where safety is a concern such as the use of large or heavy machinery, implements with moving mechanical parts, driving at high speed *etc.*, the CPU load and the soft real-time performance should be monitored continuously, and detected anomalies should be used as input to the safety system.

5.2. Evaluating if FroboMind Significantly Decreases the Resources Required to Perform Field Experiments

During the 5 day workshop the 4 man development team managed to interface FroboMind to the AMS and build the application for autonomous navigation of an orchard as well as testing the application in two trials. The trials were completed, but navigation was not completely reliable. Upon reviewing the observations at the trials and subsequent analysis of logged ROS messages it was concluded that the reliability of the in-row navigation can be increased by a few modifications to the row detecting algorithm and the navigation controller.

The workshop revealed some issues with FroboMind that need to be addressed. (1) Some of the basic components needs further development with respect to reliability, in particular interfacing to a robot platform through low level interface drivers needs to work seamlessly; (2) Interfaces between layers in the architecture need to be reviewed and properly documented; (3) The usability of FroboMind must be enhanced by a better integration with a simulation environment like Stage. These issues have not yet been solved completely, but a number of improvements have been made since the workshop took place.

As shown in Table 9 the developers worked approximately 12 hours/day each, about half the time was spent on interfacing and the other half on the orchard application. Factors like the described problems with the AMS, transport from the university to the orchard test site and changing weather conditions prolonged the work. Similarly the developer team had collectively an extensive experience in all part of the FroboMind software platform, and the AMS actuators and sensor interfaces were already installed and thoroughly tested using Mobotware, which eased the work. This makes it very difficult to compare with similar experiments, and thus to conclude if FroboMind significantly decreases the resources required to perform field experiments, but experiences from previous projects show that this is a very short time moving from a completely new software implementation on a robot to trials in the field navigating autonomously.

5.3. Evaluating Software Reuse Across Existing Projects Using FroboMind

Table 5 lists a total of 13 robots that have been interfaced to FroboMind during the past three years. With the exception of the AMS and SMR robots that are based on Mobotware they have all been constructed within the same timeframe. The use of FroboMind is not widespread though, the listed robots are spread across only 4 universities and 3 companies, most of them working together developing the field robots. In addition to the list a smaller number of robots developed by researchers, companies and students are known to either use FroboMind or have used FroboMind as a starting point for the software development.

The statistical website ohloh.net reports that the FroboMind directory structure contains 104,617 SLOC (April 2014) and the core part of the ROS platform contains 418,977 SLOC (December 2013). In this context the third column of Table 5 and the associated Table 6 listing SLOC for the components that are shared between some but not all platforms could be considered an indication of the differences in the robot software rather than the similarities.

It seems reasonable to conclude that the level of software reuse between the listed robots is high which may in part be attributed to the modular structure of the FroboMind architecture and to the flexibility that ROS provides. It is, however, also caused by the fact that the robot platforms and their current applications are fairly similar with respect to the robot software.

5.4. Lessons Learned

Choosing Ubuntu as operating system for the software platform has proven to be a great advantage. The limitations caused by not having hard real-time capabilities are counterbalanced by a number of advantages that all relates to simplicity in installation, use and maintenance and hence decreasing the time consumption from development to field trials. The ability to use the same operating system on the robot and the developers laptops speeds up the development process significantly.

The choice of ROS as middleware has had its advantages as well as drawbacks. ROS has a very steep learning curve in the beginning, but after some struggling with understanding the concepts it becomes a valuable tool. The user feedback indicates that FroboMind decreases the learning curve because it provides full working examples for simulation and execution on small robots rather than the user having to build the first test ROS setup from scratch. The ROS messaging system working across networked platforms, the ROS launch tool that ease launching of multiple ROS nodes locally and on networked computers, and the setting of parameters, as well as the many existing libraries and drivers etc. are highly valuable. The ROS tool for recording and playback of messages saves a tremendous amount of time when analyzing data and debugging. But ROS is still in a very active state of development. Experience shows that at new version releases, not all core components are fully working with the new version, and some of the new versions include substantial changes that induce unforeseen maintenance tasks. Examples are the introduction of the catkin build system and removal of the Stack concept in the ROS Groovy version. Seen in retrospect ROS was a better choice than Orocos when considering the purpose of FroboMind and the stated design goals. But it comes with the price of lacking hard real-time capabilities which is not a major problem but needs to be addressed in most applications, and the ROS code base appears to be less mature and stable than Orocos.

The FroboMind architecture design builds upon the results of the reviewed publications, especially Stanley [25], CARMEN [8] and [17] provided a significant input to the design. The architecture has undergone several revisions since the first prototype was developed in 2011. The original draft was very detailed but has been simplified significantly because experiences from using FroboMind in research projects have revealed a relationship between complexity and the willingness to accept and utilize the architecture. If the architecture is too complex, users tends to short circuit it by merging all the remaining functionality into a single or a couple of components, which makes them less suitable for reuse. It is debatable if the architecture is still too complex, this should be investigated in the future work.

An important part of the architecture is the specification of data exchanged between the components. For navigation sensors, localization and low level propulsion control and actuation this is reasonably well defined and where applicable aligned with current ROS standards. Interfacing the decision making layer and implement modules still need to be properly specified. This work should where applicable be based on existing standards within agriculture and related field work such as OpenGIS, agroXML, GPX, Isobus *etc*.

Compared to the architectures listed in Table 2 FroboMind obtains a *Yes* in all parameters which indicates that the aim of this work has been achieved. An overall performance comparison of the FroboMind software platform to solutions described in the related work is impossible though, as data relevant for comparison of the architectural and software performance has not been published.

5.5. The future of FroboMind

The primary focus in the development of FroboMind has until now been to establish a common open software platform that promotes software reuse and thus decreases the development time and resources required to perform field experiments. This objective has to a large extent been achieved and the work will continue though the application of FroboMind to new research projects, development of new components as well as improving the existing components in terms of reliability.

In addition to this a project focusing on safe behaviour of agricultural machines has been launched recently. FroboMind will be used in parts of the project, and it is expected that this will contribute significantly to the development of: (1) The safety layer which is currently limited to a deadman signal which enables the actuator controllers; (2) The decision making architecture, which currently is limited to fairly simple behaviours defined in finite state machines; (3) source code integrity through the implementation of model driven auto-generation of component interfaces and structures.

6. Conclusions

In this paper we have presented FroboMind, a software platform tailored to field robots in precision agriculture. FroboMind optimizes field robot software development in terms of code reuse between different research projects. At the current stage FroboMind has been ported to 2 different 4-wheeled tractors, 7 different configurations of tracked field robots and 4 differentially steered robots. Examples of current FroboMind applications are autonomous crop scouting, precision spraying, mechanical weeding, grass cutting, humanitarian demining and land surveying.

In an experiment evaluating the performance of FroboMind in robot systems with near real-time requirements it was concluded that FroboMind's soft real-time execution is sufficient for typical precision agriculture tasks such as autonomous navigation of a predefined route. In any application where safety is a concern, the CPU load and the soft real-time performance should be monitored continuously, and detected anomalies should be used as input to the safety system.

To test whether FroboMind decreases the development time and resources required to perform precision agriculture field experiments, an experiment was performed porting FroboMind to a new field robot and applying this robot to autonomous navigation in an orchard using local sensors. This was achieved by 4 developers during a 5 day workshop, which is a very short time moving from a completely new software implementation on a robot to trials in the field navigating autonomously.

The software reuse across projects has been assessed using available data from existing projects and robot platforms. It was concluded that the level of software reuse between the listed robots is high which may in part be attributed to the modular structure of the FroboMind architecture, the flexibility that ROS provides and the fact that the robot platforms and their current applications are fairly similar with respect to the robot software.

It is concluded that FroboMind is usable in practical field robot applications by research groups and other stakeholders. Further development of FroboMind continue through new research projects where the current activities focus on the decision making structure and autonomous safe behaviour. The FroboMind software has been released as open-source at the FroboMind website [32] for others to build upon.

Acknowledgements

This work is linked to and partially funded by the ERA-Net ICT-AGRI project: Grassbots, the Danish Ministry for Food, Agriculture and Fisheries project: FruitGrowth, and the Danish National Advanced Technology Foundation project: The Intelligent Sprayer Boom. The authors wish to thank Claes D. Jaeger, Hans W. Griepentrog, Ulrik P. Schultz and Mikkel K. Larsen for supporting this work.

Authors Contributions

Kjeld Jensen: Main author, main architect of FroboMind, developed FroboMind components, designed the experiments and used FroboMind in research projects; Morten Larsen, Søren H. Nielsen, Leon B. Larsen and Kent S. Olsen: Contributed to the FroboMind design and experiments, developed FroboMind components, used FroboMind in research projects and gave important feedback on drafts; Rasmus N. Jørgensen: Used FroboMind in research projects and gave important feedback on drafts.

Conflict of Interest

The authors declare no conflict of interest.

References

1. Pedersen, S.M.; Fountas, S.; Have, H.; Blackmore, B.S. Agricultural robots—System analysis and economic feasibility. *Precis. Agric.* **2006**, *7*, 295–308.
2. Green, O.; Schmidt, T.; Pietrzkowski, R.P.; Jensen, K.; Larsen, M.; Jørgensen, R.N. *Commercial Autonomous Agricultural Platform*; Kongskilde Robotti: Soroe, Denmark, 2014.
3. Bakker, T.; van Asselt, K.; Bontsema, J.; Muller, J.; van Straten, G. Systematic design of an autonomous platform for robotic weeding. *J. Terramechanics* **2010**, *47*, 63–73.
4. Ruckelshausen, A.; Biber, P.; Dorna, M.; Gremmes, H.; Klose, R.; Linz, A.; Rahe, R.; Resch, R.; Thiel, M.; Trautz, D.; Weiss, U. BoniRob: An Autonomous Field Robot Platform for Individual Plant Phenotyping. Avaliable online: https://my.hs-osnabrueck.de/ecs/fileadmin/groups/156/Veroeffentlichungen/2009-JIAC-BoniRob.pdf (accessed on 12 May 2014).

5. Jørgensen, R.; Sørensen, C.; Pedersen, J.; Havn, I.; Jensen, K.and Søgaard, H.; L.B., S. Hortibot: A system design of a robotic tool carrier for high-tech plant nursing. *CIGR E J.* **2007**, *9*, ATOE 07 006.

6. Blackmore, B.S.; Griepentrog, H.W.; Nielsen, H.; Nørremark, M.; Resting-Jeppersen, J. Development of a deterministic autonomous tractor. In Proceedings of the 2004 CIGR Olympics of Agricultural Engineering, Beijing, China, 11–14 October 2004.

7. Neisser, U. *Cognitive Psychology*; Meredith Publishing: New York, NY, USA, 1967.

8. Montemerlo, M.; Roy, N.; Thrun, S. Perspectives on standardization in mobile robot programming: The Carnegie Mellon navigation (CARMEN) toolkit. In Proceedings of the IEEE/RSJ International Conference on Intelligent Robots and Systems, Las Vegas, NV, USA, 27–30 October 2003.

9. Nesnas, I.A.D.; Wright, A.; Bajracharya, M.; Simmons, R.; Estlin, T. CLARAty and challenges of developing interoperable robotic software. InProceedings of the 2003 IEEE/RSJ International Conference on Intelligent Robots and Systems, Las Vegas, NV, USA, 27–30 October 2003; Volumes 1–4, pp. 2428–2435.

10. Pivtoraiko, M.; Nesnas, I.A.; Nayar, H.D. A Reusable Software Framework for Rover Motion Controls. Available online http://robotics.estec.esa.int/i-SAIRAS/isairas2008/Proceedings/SESSION%2011/m086-Pivtoraiko.pdf (accessed on 12 May 2014).

11. Cepeda, J.; Chaimowicz, L.; Soto, R. Exploring Microsoft Robotics Studio as a Mechanism for Service-Oriented Robotics. In Proceedings of the Robotics Symposium and Intelligent Robotic Meeting (LARS), Latin American, Sao Bernardo do Campo, Brazil, 23–28 October 2010; pp. 7–12.

12. Makarenko, A.; Brooks, A.; Kaupp, T. Orca: Components for Robotics. In Proceedings of the Workshop on Robotic Standardization, IEEE/RSJ International Conference on Intelligent Robots and Systems (IROS 2006), Beijing, China, 9–15 October 2006.

13. Henning, M. A New Approach to Object-Oriented Middleware. *IEEE Internet Comput.* **2004**, *8*, 66–75.

14. Bruyninckx, H. Open robot control software: The OROCOS project. In Proceedings of the IEEE International Conference on Robotics and Automation, Seoul, Korea, 21–26 May 2001.

15. Gerkey, B.P.; Vaughan, R.T.; Sukhatme, G.S.; Stoy, K.; Mataric, M.J.; Howard, A. Most valuable player: A robot device server for distributed control. In Proceedings of the IEEE/RSJ International Conference on Intelligent Robots and Systems, Maui, HI, USA, 29 October–3 November 2001.

16. Quigley, M.; Conley, K.; Gerkey, B.P.; Faust, J.; Foote, T.; Leibs, J.; Wheeler, R.; Ng, A.Y. ROS: An open-source Robot Operating System. In Proceedings of the ICRA Workshop on Open Source Software, Kobe, Japan, 12–17 May 2009.

17. Vázquez-martín, R.; Martinez, J.; Toro, J.C.; Núñez, P. A Software Control Architecture based on Active Perception for Mobile Robotics 1. Available online: http://www.researchgate.net/publication/228638002ASoftwareControlArchitecturebasedonActivePerceptionforMobileRobotics/file/5046351f943eef19bb.pdf (accessed on 12 May 2014).

18. Brooks, R.A. A Robust Layered Control-system For A Mobile Robot. *IEEE J. Robot. Autom.* **1986**, *2*, 14–23.

19. Patricio Nebot, J.T.S.; Martinez, R.J. A New HLA-Based Distributed Control Architecture for Agricultural Teams of Robots in Hybrid Applications with Real and Simulated Devices or Environments. *Sensors* **2011**, *11*, 4385–4400.

20. Garcia-Perez, L.; Garcia-Alegre, M.C.; Ribeiro, A.; Guinea, D. An agent of behaviour architecture for unmanned control of a farming vehicle. *Comput. Electron. Agric.* **2008**, *60*, 39–48.

21. Kuhnert, K.D. Software Architecture of the Autonomous Mobile Outdoor Robot AMOR. In Proceedings of the 2008 IEEE Intelligent Vehicles Symposium, Eindhoven, The Netherlands, 4–6 June 2008.

22. Beck, A.B.; Andersen, N.A.; Andersen, J.C.; Ravn, O. MobotWare—A plug-in based framework for mobile robots. In Proceedings of the 7th IFAC Symposium on Intelligent Autonomous Vehicles, Lecce, Italy, 6–8 September 2010; Volume 7, pp. 127–132.

23. Simon Blackmore, S.F.; Have, H. Proposed System Architecture to Enable Behavioral Control of an Autonomous Tractor. Automation Technology for Off-Road Equipment. In Proceedings of the American Society of Agricultural and Biological Engineers (ASABE), Chicago, IL, USA, 26–27 July 2002. pp. 13–23.

24. Fountas, S.; Blackmore, B.S.; Vougioukas, S.; Tang, L.; Sørensen, C.G.; Jørgensen, R. Decomposition of Agricultural tasks into Robotic Behaviours. *Agric. Eng. Int. CIGR Ejournal* **2007**, *IX*, Manuscript PM 07 006.

25. Thrun, S.; Montemerlo, M.; Dahlkamp, H.; Stavens, D.; Aron, A.; Diebel, J.; Fong, P.; Gale, J.; Halpenny, M.; Hoffmann, G.; *et al.* Stanley: The robot that won the DARPA Grand Challenge. *J. Field Robot.* **2006**, *23*, 661–692.

26. Dozio, L.; Mantegazza, P. Linux Real Time Application Interface (RTAI) in low cost high performance motion control. In Proceedings of the Motion Control 2003, a Conference of ANIPLA, Associazione Nazionale Italiana per l'Automazione (National Italian Association for Automation), Milano, Italy, 27–28 March 2003.

27. Carsten Emde, T.G. *Quality Assessment of Real-Time Linux*; Technical Report, Open Source Automation Development Lab: Stuttgart, Germany, 2011.

28. McKenney, P.E. *'Real Time' vs. 'Real Fast': How to Choose?* Technical Report, IBM Linux Technology Center: New York, NY, USA, 2008.

29. Leonard, J.; Durrant-Whyte, H. Mobile robot localization by tracking geometric beacons. *IEEE Trans. Robot. Autom.* **1991**, *7*, 376–382.

30. Stuart Russell, P.N. *Artificial Intelligence: A Modern Approach*, 3rd ed.; Prentice Hall: Upper Saddle River, NJ, USA, 2010.

31. Schacter, D. *Psychology*; Worth Publishers: London, UK, 2011.

32. FroboMind Website. Available online: http://www.frobomind.org accessed on (5 May 2014).

33. Jaeger-Hansen, C.; Jensen, K.; Larsen, M.; Nielsen, S.; Griepentrog, H.; Jørgensen, R. Evaluating the portability of the FroboMind architecture to new field robot platforms. In Proceedings of the 9th Europaean Confernce on Precision Agriculture, Lleida, Spain, 7–11 July 2013.

34. Jensen, K.; Laursen, M.S.; Midtiby, H.; Jørgensen, R.N. Autonomous Precision Spraying Trials Using a Novel Cell Spray Implement Mounted on an Armadillo Tool Carrier. In Proceedings of the XXXV CIOSTA & CIGR V Conference, Billund, Denmark, 3–5 July 2013.

35. Larsen, L.B.; Olsen, K.S.; Ahrenkiel, L.; Jensen, K. Extracurricular Activities Targeted towards Increasing the Number of Engineers Working in the Field of Precision Agriculture. In Proceedings of the XXXV CIOSTA & CIGR V Conference, Billund, Denmark, 3–5 July 2013.

36. Barawid, O.C.; Mizushima, A.; Ishii, K.; Noguchi, N. Development of an autonomous navigation system using a two-dimensional laser scanner in an orchard application. *Biosyst. Eng.* **2007**, *96*, 139–149.

37. Blas, R. Fault-Tolerant Vision for Vehicle Guidance in Agriculture. Ph.D. Thesis, Technical University of Denmark, Lyngby, Denmark, 2010.

38. Reske-Nielsen, A.; Mejnertsen, A.; Andersen, N.A.; Ravn, O.; Nørremark, M.; Griepentrog, H.W. Multilayer Controller for Outdoor Vehicle. In Proceedings of the Automation Technology for Off-Road Equipment (ATOE), Bonn, Germany, 1–2 September 2006; pp. 41–49.

39. Griepentrog, H.; Andersen, N.; Andersen, J.; Blanke, M.; Heinemann, O.; Madsen, T.; Nielsen, J.; Pedersen, S.; Ravn, O.; Wulfsohn, D. Safe and Reliable: Further Development of a Field Robot. Available online: www.staff.kvl.dk/ hwg/pdf/papers/Griepentrog2009-7ECPA.pdf (accessed on 12 Mary 2014).

5

DOF Decoupling Task Graph Model: Reducing the Complexity of Touch-Based Active Sensing

Niccoló Tosi [1,2,*], Olivier David [2] and Herman Bruyninckx [1,3]

[1] Department of Mechanical Engineering, KU Leuven, Celestijnenlaan 300, 3001 Heverlee, Belgium;
  E-Mail: niccolo.tosi@kuleuven.be (N.T.); herman.bruyninckx@mech.kuleuven.be (H.B.)
[2] CEA, LIST, Interactive Robotics Laboratory, PC 178, 91191 Gif sur Yvette Cedex, France;
  E-Mail: odavid@cea.fr
[3] Department of Mechanical Engineering, Eindhoven University of Technology, Eindhoven,
  The Netherlands

* Author to whom correspondence should be addressed; E-Mail: niccolo.tosi@kuleuven.be

Academic Editors: Nicola Bellotto, Nick Hawes, Mohan Sridharan and Daniele Nardi

---

Abstract: This article presents: (i) a formal, generic model for active sensing tasks; (ii) the insight that active sensing actions can very often be searched on less than six-dimensional configuration spaces (bringing an exponential reduction in the computational costs involved in the search); (iii) an algorithm for selecting actions explicitly trading off information gain, execution time and computational cost; and (iv) experimental results of touch-based localization in an industrial setting. Generalizing from prior work, the formal model represents an active sensing task by six primitives: configuration space, information space, object model, action space, inference scheme and action-selection scheme; prior work applications conform to the model as illustrated by four concrete examples. On top of the mentioned primitives, the task graph is then introduced as the relationship to represent an active sensing task as a sequence of low-complexity actions defined over different configuration spaces of the object. The presented `act-reason` algorithm is an action selection scheme to maximize the expected information gain of each action, explicitly constraining the time allocated to compute and execute the actions. The experimental contributions include localization of objects with: (1) a force-controlled robot equipped with a spherical touch probe; (2) a geometric complexity of the to-be-localized objects up to industrial relevance; (3) an initial uncertainty of (0.4 m, 0.4 m, 2π); and (4) a configuration of `act-reason` to constrain the allocated time to compute and execute the next action as a function of the current uncertainty. Localization is accomplished when the probability mass within a 5-mm tolerance reaches a specified threshold of 80%. Four objects are localized with final {mean; standard-deviation} error spanning from {0.0043 m; 0.0034 m} to {0.0073 m; 0.0048 m}.

Keywords: active sensing; localization; tactile sensors; information gain; entropy; decision making; reasoning

---

1. Introduction

1.1. Motivation and Use Cases

Typical robotics applications, e.g., bin picking, assembly or object manipulation, require the robot to be able to place its end-effector in a given location within a specified tolerance with respect to the objects on which it operates. A reliable 3D model of the environment, in robot-centric coordinates, becomes a fundamental building block for such operations. Vital to enhance the reliability of the 3D model, scene calibration is the act of estimating the pose (*i.e.*, the position and orientation) of the objects in the environment.

Built on a CAD representation of the scene objects, the virtual model may be exploited, for instance, to assist the human operator, while performing tele-manipulation tasks or to plan actions in cluttered environments, avoiding undesired contacts [1]. In order to perform calibration, different forms of sensing can be adopted, including vision, laser, ultrasound and touch. Our research focuses on touch sensing, and in particular, it has been inspired by the practical need of performing model calibration for tele-manipulation applications during tunnel-boring operations [2]. Specifically, the experimental results presented in this work are obtained with a force-torque sensor, which is the most suitable touch probe for such harsh environments. Therefore, this work presents modeling representations developed in such a context. Nevertheless, their formulation is purposely generic, so that they can be applied to any active-sensing task.

In general, scene calibration may become non-trivial and potentially time-consuming. In particular, the decision making about where to sense next requires significant computational resources in order to identify the most informative action to perform. This is due to the cumbersomeness of reasoning in the belief state space of the object locations. In the context of touch-based active sensing for probabilistic localization, information-gain metrics, such as the weighted covariance matrix [3], probability entropy [4–7] or the Kullback–Leibler divergence [8], have been adopted as reward functions to select actions. However, touch localization typically presents non-Gaussian conditions, since the posterior distribution on the pose of the object is often multi-modal due to the local observation collected (e.g., contact point and normal vector of the touched surface). Under non-Gaussian conditions, decision-making schemes, such as a partially-observable Markov decision process (POMDP), suffer from the curse of dimensionality and do not scale well when the initial uncertainty increases.

Driven by the need to perform object localization in industrial settings, this work focuses on reducing the time to compute, execute and process active sensing tasks, dealing with objects of industrial complexity.

1.2. Contributions

The main contributions of this work are summarized below.

- The active sensing task model is defined generalizing from related work in the literature. Specifically, an active-sensing task is the composition of six modeling primitives: configuration space, information space, object model, action space, inference scheme and action-selection scheme. Concrete literature examples conforming to the model are provided.

- The `act-reason` algorithm is presented as an action-selection scheme that is designed to explicitly trade off information gain (which is the traditional "reward" of any active sensing task) with the execution time of the action and the time to compute what action to select (which are "cost" factors that are seldom taken into account in the literature). Specifically, the time allocated to execute and evaluate the next action is offered as a configurable design parameter to the robot programmer.

- The DOF decoupling task graph is introduced as the model that allows the task programmers to represent different strategies in the design of localization tasks, as (partially ordered) sequences of active-sensing subtasks with the lowest possible complexity in the above-mentioned composition of modeling primitives. In contrast to the literature (where all subtasks are defined in the same, highest-dimensional (typically six-dimensional) configuration and action spaces), the possibility to select actions in lower-dimensional spaces (typically two- and three-dimensional actions to find the position and orientation of an object on a planar support surface) allows one to significantly minimize the complexity of the computations to select actions, hence also speeding up the overall execution of active localization tasks.

- Experimental tests are carried out with a Staubli RX90 robot. A series of objects up to industrial complexity are localized using touch-based sensing with an industrial force sensor. A 3DOF localization with (0.4 m, 0.4 m, 360 degree) initial uncertainty is carried out applying an empirical task graph (defined in Section 4) in order to reduce task complexity. In these experiments, object localization is always performed in less than 120 s concentrating 80% of the probability mass within a specified feature on the object. The `act-reason` algorithm is configured such that the time allocated to computing the next action is a function of the current uncertainty about the pose of the object; this results in time efficiency improvements over random and entropy-based schemes.

2. Definition of Key Concepts

This work focuses on using force sensing as a reliable means to explore the robot environment. In this context, a sensing primitive is comprised of a robot motion and the force-torque signals measured by a sensor coupled with the robot end effector. Generalizing from this concrete case leads to the following definitions at the conceptual level.

Task: any controlled motion of the robot interpreting information collected through its sensory system in order to realize a planned change in the world.

More specifically, control is responsible for realizing the planned change, while sensing collects information about the actual state of the world. Planning selects sensing and controls that are expected to realize desired changes in the world in "the most optimal way". In general, a task is a composition of other tasks of lower "levels of abstraction". At the conceptual level, we deliberately avoid the definition of concepts, like "action" or "primitive task", since they only become relevant as soon as one introduces concrete formal representations (in other words, "models") of tasks at a defined level of abstraction.

Active sensing task: a task whose desired outcome is to gain information about the pose of objects in the world. In the context of object localization, one has to make concrete choices of models to represent the world, *i.e.*, to describe where the objects are in the environment. The choice of world representation influences the choice of the models to represent the controller, the sensing and the planning. Section 3.1 presents a model-level definition of active-sensing tasks in the context of object localization.

Configuration space of a system: a set of the possible states of that system. In general, this definition applies to all of the systems related to a task, e.g., world, control, sensing and planning. Since this work focuses on object localization, when used without an explicit context, the term "configuration space" refers to the set of possible poses that the object can have in the world.

Degrees of freedom (DOFs) of a system: the number of independent parameters to represent the system's configuration space. The experiments in this work are carried out in configuration spaces with two and three DOFs; but, all its theoretical contributions about the localization of objects hold in any relevant configuration space, including the general full six-dimensional one.

Complexity: the amount of resources required to describe a configuration space. Since planning, sensing and control are all working with the representation of the world, the complexity of the configuration space of the pose of the objects has a direct influence on the complexity of the representations and algorithms for planning, sensing and control. In general, complexity metrics are task dependent, since different systems and processes exploit different forms of resources. For instance, let us represent a configuration space discretizing it with a uniform grid. The number of necessary nodes N can be used as the complexity metric for the configuration space. Specifically, N depends on the number of DOFs (n_{DOFs}) composing the configuration space, on the grid resolution res(*i.e.*, the distance between two adjacent nodes) and on its range (*i.e.*, the difference between the maximum and the minimum value for each DOF). Hence, the complexity of the configuration space is that expressed in Equation (1).

$$N = \prod_{i=1}^{n_{DOFs}} \frac{range_i}{res_i} \qquad (1)$$

The experimental contributions of this work are comprised of touch-based localizations where the world is represented by a geometric model composed of polygonal objects. Their configuration space, *i.e.*, their pose in the world, will be expressed by DOFs defined with respect to the robot frame of reference. In such a context, the model-level definition of an action is here provided.

Action: a controlled robot trajectory, which provides observations of the pose of the objects in the world, in a robot-centric reference frame. The quality and quantity of the observations collected during an action depend on the control and the sensing apparatus.

Reward: a function to be maximized during action selection. In active sensing, the reward typically corresponds to the information gain.

3. Active Sensing Task Model

In this section, we introduce a model to represent active-sensing tasks formally, using the following six modeling primitives: configuration space, information space, object model, action space, inference scheme and action-selection scheme. The presented model is formulated generalizing from previous touch-based localization examples, and four concrete instances from the literature are provided. However, the proposed representation is designed to be fully generic, and hence, it is relevant to tasks that are not restricted to touch-based sensing only.

3.1. Active Sensing Task Model: Primitive Concepts

An active sensing task can be represented by a set of instances of the following primitive concepts:

- Configuration space X: the model to describe where the objects are in the world with respect to the robot reference frame.
- Information space $P(X_t|a_{0:t}, z_{0:t})$: the probability distribution that encodes the knowledge about the configuration space at time t, given the sequence of actions $a_{0:t}$ and observations $z_{0:t}$.
- Object model: a formal representation of the geometry of the object.
- Action space A: the set of sensori-motor controlled actions the robot can execute.
- Inference scheme: an algorithm to compute $P(X_t|a_{0:t}, z_{0:t})$.
- Action-selection scheme: an algorithm to make the decision for which action from the action space to execute next.

These six primitives are correlated (e.g., the action space depends on the object model, in general) and influence the complexity of the active-sensing task. Concretely, such complexity can be measured by the time required to update the posterior distribution $P(X_t|a_{0:t}, z_{0:t})$ or to select the next action from A. In practice, the latter typically turns out to be the most time-consuming operation [4,8]. The action space may be composed by actions executed exploiting different sensing capabilities of the robot, and they are not restricted to touch sensing only.

The aforementioned model is created to generalize from previous active-sensing works. Here, four examples conforming to the model are reported.

1. An infrared sensing application where a plate is localized in 2DOFs [3].
2. A force sensing application where a cube is localized in 3DOFs [9].
3. A force sensing application where objects are localized in 6DOFs [10].
4. A tactile sensing application where objects are localized in 3DOFs for grasping [4].

Graphically represented in Figure 1, they are summarized in Table 1 in terms of the six primitives.

Table 1. Prior examples of touch-based active sensing described with the six primitive concepts of the active sensing task model: configuration space, information space, object model, action space, inference scheme and action-selection scheme. POMDP, partially-observable Markov decision process.

Ex.	Ref.	Configuration Space	Information Space	Object Model	Action Space	Inference Scheme	Action Selection Scheme
1	[3]	2DOF pose of the target position to drill the hole	parametric posterior distribution	analytical model (rectangle)	infrared edge detection	extended Kalman filter	weighted covariance trace min.
2	[9]	3DOF pose of the object	discretized posterior distribution	polygonal mesh	depth-sampling with force sensor	grid method	not declared
3	[10]	6DOF pose of the object	discretized posterior distribution	polygonal mesh	depth-sampling with force sensor	scaling series	random selection
4	[4]	3DOF pose of the object	discretized posterior distribution	polygonal mesh	grasping and sweeping motions	histogram filter	POMDP with entropy min.

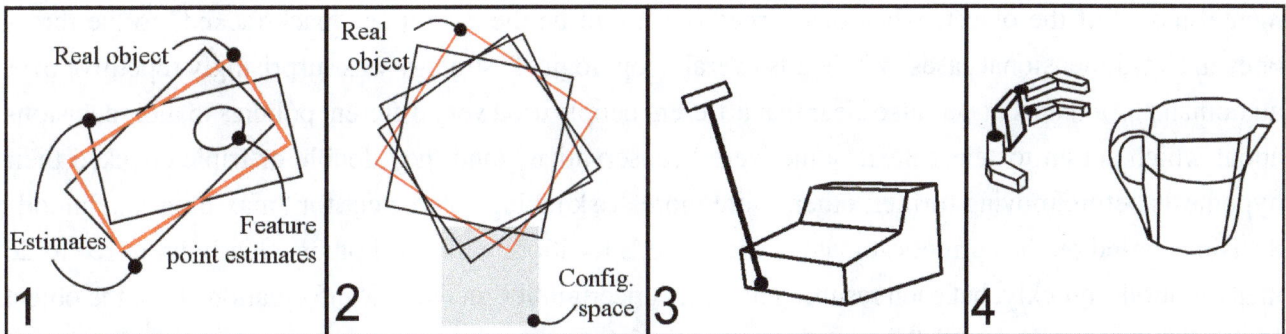

Figure 1. Graphical illustration of literature examples of touch-based active sensing tasks conforming to the active sensing task model. Image numbers refer to Table 2. Images 1 to 4 are inspired from [3], [9], [10] and [4], respectively.

4. DOF Decoupling Task Graph Model

4.1. Motivation

In Bayesian touch-based localization applications, the pose of an object is estimated by updating a belief over the configuration space, conditioned on the obtained measurements, that is an estimate of the position of the touch probe at the moment the ongoing motion is stopped because a pre-defined threshold on the force has been exceeded. Typically, such a probability distribution becomes multi-modal, due to the local nature of the available information, e.g., taking the form of a contact point and a normal vector. Under these conditions, optimal and computationally-efficient parametric Bayesian filtering models, such as the Kalman filter are not suited to capture the different modes of the probability distribution.

Hence, non-parametric inference schemes relying on a discretization of the probability distribution must be adopted, such as grid-based methods [8,9] or particle filter [11,12]. Unfortunately, the computational complexity of these inference schemes linearly increases with that of the configuration space. Moreover, the complexity of the action selection is exponential in the DOFs of the configuration space and in the time horizon of the decision making. As a matter of fact, this sets an upper bound on the size of the problems that can be treated online, since the available resources are always finite. This curse of dimensionality is a strong driver to focus on the reduction of the complexity of active sensing tasks.

In a previous work, we studied human behavior during touch-based localization [13]. A test was carried out with 30 subjects asked to localize a solid object in a structured environment. They showed common behavioral patterns, such as dividing the task into a sequence of lower complexity problems, proving remarkable time efficiency in accomplishing the task compared to state-of-the-art robotics examples of similar applications. The key insight extracted from these experiments was that humans use "planar" motions (that is, only needing a two-dimensional configuration space) as long as they have not yet touched the object; then they switch to a three-dimensional configuration space, in order to be able to represent the position and orientation in a plane, of the ("polygonal hull" of the) object they had touched before; and only after that position and orientation have been identified with "sufficient" accuracy, they start thinking in a six-dimensional configuration space, to find specific geometric features on the object, such as grooves or corners. Several times, the humans were driven by finding out whether one particular hypothesis of such a geometric feature indeed corresponded to the one they had in their mental model of the object; when that turned out not to be the case, they "backtracked" to the three- or even two-dimensional cases. While this overall "topological" strategy was surprisingly repetitive over all human participants, it was also clear that different people used very different policies in their decisions about which action to select next: some were "conservative", and they double or triple checked their hypothesis before moving further; others were more "risk taking", and went for "maximum likelihood" decisions based on the current estimate of the object's location; while still others clearly preferred to do many motions, quickly, but each resulting in larger uncertainties on the new information about the object location than the more carefully-executed motions of the conservative group.

Inspired by this result, approaches were sought to also allow robots to divide a localization application into a series of subtask adapting the problem complexity at run time; the primitives that are available for this adaptation are: execution time, information gain per action, time to compute the next action and the choice of the configuration space over which the search is being conducted. A formal representation of these adaptation possibilities would allow the robot to reason itself about how to tackle tasks with higher initial uncertainty, performing the action selection online. In this work, "online" means that the reasoning about which action to execute does not take longer than the actual action execution itself. Indeed, if reasoning takes longer, it would make more sense to let the robot just execute random motions, which easily do better than not moving at all while computing how to move "best".

4.2. Model Formulation: Task Graph and Policy

Backed by the results of the human experiment and building upon [14], it became clear that a more extended task representation formulation was required, to allow task programmers to explicitly represent a robot-localization application: (i) in a modular fashion; (ii) with each module specified in a configuration space of the "lowest possible" dimension; and (iii) using one of several possible action selection strategies. To this end, the task graph is added to the task model as a new modeling primitive, with the explicit purpose of representing the relationship that connects the just-mentioned pieces of information to a sequence of subtasks: each such subtask is defined (possibly recursively) as an active sensing module, with concrete instances for all of its six task model primitives and with a concrete choice of configuration space dimension and a concrete choice of reasoning strategy. This formal representation then contains all of the information to reduce, at run-time, the computational complexity of estimation and action-selection.

Task graph: an oriented graph representing a localization application in which each node corresponds to an active-sensing task defined in its specific configuration space and with a specific search strategy. The task graph allows us to encode the sequential order in which subtasks can be executed. Arcs encode transitions between subtasks, which are triggered by events, for instance: a sensor signal reaching some specified threshold, a change in the level of uncertainty or a human-specified occurrence.

Policy: a path through a task graph, connecting subtasks and transitions. Different policies may be defined in a task graph. For instance, this could be done further to an empirical observation, as proposed in the experimental section of this work, or by learning. In order to select a policy, graph-search techniques from the literature can be applied, e.g., random selection, breadth-first or depth-first. Hence, as proposed in [14], an active sensing localization task requires decision making at two different levels:

1. to select a policy, *i.e.*, the sequence of subtasks to execute;
2. to select an action within each subtask.

In general, the structure of the task graph is application dependent. For the sake of clarity, Figure 2 illustrates an example of a localization application described as a task graph with subtasks defined over three different configuration spaces: C1, C2 and C3. Three policies are available in the presented case, depicted with arrows of different patterns. Subtasks are represented as circles. It is important to note that several subtasks may be defined on the same configuration space, e.g., with different information space, object model, action space, inference and action-selection scheme. Transition events are depicted with a filled circle.

In this work, the DOF decoupling task graph model is exploited for touch-based active sensing applications. Policy generation and selection is done empirically, further to the human experiment observation. Building upon our previous work [15], the following section introduces a new action-selection algorithm, which is implemented on a practical case study in Section 7.3.

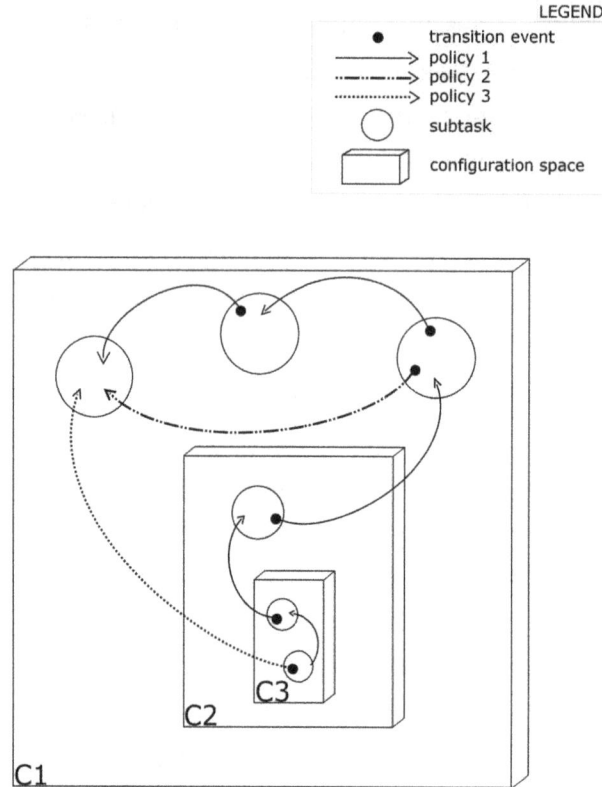

Figure 2. Example of task graph. Graphically, solid rectangles represent configuration spaces and circles represent subtasks. In this specific case, the task is defined over three configuration spaces, C1, C2 and C3, presenting three, one and two tasks, respectively. The localization is initialized in configuration space C3. Three policies are depicted with arrows of different patterns, each of them representing a possible way through the tasks defined over C1, C2 and C3.

5. The `act-reason` Algorithm: Making Decisions over a Flexible Action Space

To the best of the authors' knowledge, previous works on active sensing made decisions about where to sense next, maximizing some reward function choosing from a pre-defined action space composed of a set of sensing motions either specified by a human demonstrator or automatically generated off-line [4,5,8]. This resulted in an implicit choice on the complexity of the action space, hence also increasing the complexity of the action selection. In practice, this often caused the robot to devote a significant percentage of the running time to reasoning instead of acting.

5.1. Motivation

In the framework of the active sensing task model, there is the need to make the choice on the complexity of the action space explicit, thus allowing the robot to adjust the action-selection complexity at run time. With respect to the state-of-the-art, we introduce a new action-selection scheme named `act-reason`, which allows the robot to find a solution to the problem of where to sense next, constraining the time t_{alloc} spent for the next action, including reasoning and motion execution. Therefore, the complexity of the action space depends on t_{alloc} and the time required to reason about each action. In `act-reason`, t_{alloc} is a design parameter that sets the robot behavior from info-gathering (increasing t_{alloc}) to motion-oriented (reducing t_{alloc}). In particular, this makes the action space flexible, so the complexity of the decision making can vary at run time.

5.2. Algorithm

Previous related works [5,8,16,17] tackled the problem of choosing the next best action a^* by first generating a complete set of candidate actions $\{a\}$, then selecting the one that maximizes some specified reward function $r(.)$:

$$a^* = \arg\max_a E[r(a)] \tag{2}$$

Often, r corresponds to the information gain [3,7,14] or some task-oriented metric, such as the probability of grasping success, as proposed in [18]. However, the `act-reason` algorithm is independent of the chosen reward function, so it may apply to a generic robot task. For each sensing action a, the proposed scheme assumes the robot to be able to:

- generate the associated motion;
- evaluate the expected reward;
- estimate the execution time t_{exec};
- estimate the evaluation time t_{eval}.

Intuitively, the action space is continuously sampled to estimate the expected reward that can be achieved within a preset total allocated time t_{alloc}. At each time step, the current best solution a^* is stored. The t_{alloc} parameter bounds the time spent to evaluate and execute the next action. Figure 3 graphically represents the action selection at different time steps: the starting time t_0, a generic instant t_i and t_{final} when a^* is executed. In the illustrated case, an open-loop action is available, so a^* is initialized to a_{OL}. As time passes, a^* is updated, until t.elapsed() intersects $t_{exec}(a^*)$, and the current best action is executed.

The `act-reason` algorithm is presented in pseudo-code in Algorithm 1. Each time the robot needs to make a decision, the best solution a^* is initialized to "stay still", unless an open-loop action is available, e.g., a user-defined action (Lines 2–8). Then, the timer starts (Line 9). While the execution time of the current best solution is smaller than the available time ($t_{alloc} - t$.elapsed()), candidate actions are generated and evaluated, updating the best solution if the expected reward is greater than the current one (Lines 16–18). As the available time intercepts the estimated execution time of the current best solution, the reasoning is over, and a^* is performed. The condition at Line 14 ensures that the next candidate to evaluate respects the time constraint.

A practical implementation of the `act-reason` algorithm is presented in Section 7.3, where t_{alloc} is configured to minimize the execution time during a touch-based localization task.

Algorithm 1 Action selection with `act-reason`.

1: set t_{alloc}
2: **if** a_{OL} is available **then**
3:      $a^* = a_{OL}$
4:      $E[r^*] = E[r(a_{OL})]$
5: **else**
6:      $a^* =$ "stay still"
7:      $E[r^*] = 0$
8: **end if**
9: $t.start()$
10: $i = 0$
11: **while** $t_{exec}(a^*) < t_{alloc} - t.elapsed()$ **do**
12:      $i := i + 1$
13:      $a_i = $ generateCandidateAction()
14:      **if** $t_{exec}(a_i) + t_{eval}(a_i) < t_{alloc} - t.elapsed()$ **then**
15:          compute $E[r(a_i)]$
16:          **if** $E[r(a_i)] > E[r(a^*)]$ **then**
17:              $a^* = a_i$
18:          **end if**
19:      **end if**
20: **end while**
21: **if** $a^* ==$ "stay still" **then**
22:      $t_{alloc} := 2\, t_{alloc}$
23:      go back to line 2
24: **end if**
25: execute a^*

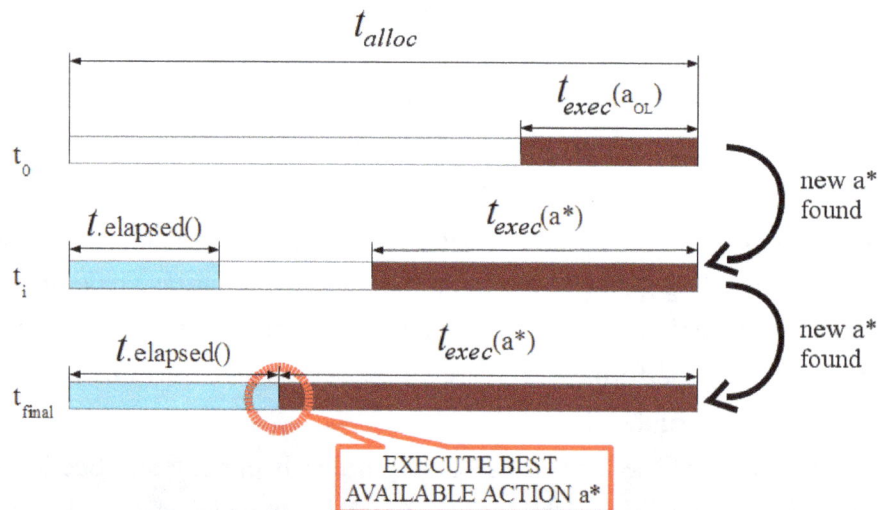

Figure 3. Action-selection time line with `act-reason` decision making. New candidate actions are generated and evaluated until the available time equals that required to execute the current best solution.

6. Robot Setup and Measurement Model

This work presents robotic applications of the DOF decoupling task graph model to touch-based localization. Specifically, the robot has to estimate the pose $\mathbf{x}$ of an object modeled as a polygonal mesh composed of faces $\{f_i\}$ and their associated normal vectors $\{n_i\}$. A single $< f_i, n_i >$ tuple will be referred to as patch. The pose of the object is represented by the state vector $\mathbf{x}$, which collects its position and roll-pitch-yaw orientation with respect to the robot frame of reference $\mathbf{R}$. In a Bayesian framework, the information on the pose of the object is encoded in the joint probability distribution over $\mathbf{x}$ conditioned on the measurement z: $P(\mathbf{x}|z)$. This section describes the robot set up and the measurement models adopted.

6.1. Contact Detection Model

The robot is equipped with a spherical end-effector coupled with a force-torque sensor (as shown in Figure 4), and it acquires information about $\mathbf{x}$ both when moving in free space and when in contact with the object, assuming that interactions only take place with its end-effector. In this context, an action consists of the end-effector following a trajectory τ, sweeping a volume $\psi(\tau)$. Concretely, τ is a collection of end-effector frames $\{\phi\}$ used as a requirement for the robot controller. When a contact is detected, the motion is stopped with the end-effector in frame ϕ_c. In case of no-contact, the whole trajectory is executed with the end-effector reaching pose ϕ_{end}. The sensory apparatus is comprised of a force-torque sensor used as a collision detector setting an amplitude threshold on the measured forces. Specifically, it detects whether the end-effector is in contact with the object (C) or is moving in free space (NC):

$$z_{CD} = \begin{cases} C, \text{ if the resultant force reaches the threshold} \\ NC, \text{ otherwise} \end{cases} \tag{3}$$

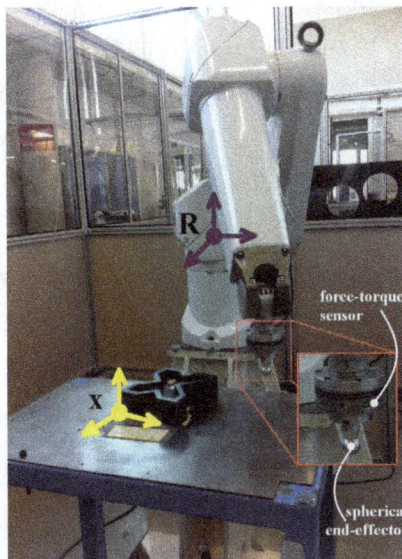

Figure 4. Robot setup used for the experimental test. The pose of the object $\mathbf{x}$ is represented with respect to the robot reference frame R.

When the end-effector touches the object, the point of contact z_P and the normal vector z_n of the touched point can be estimated following the procedure described by Kitagaki *et al.* [19], under the assumption of frictionless contact. We will refer to this touch measurement as z_{touch}:

$$z_{touch} = \{z_P, z_n\} \tag{4}$$

As in [4], in order to define the observation models, we consider two possible cases: (i) the robot reaches the end of the trajectory; and (ii) the robot makes contact with the object stopping in ϕ_c.

6.2. The Robot Reaches the End of the Trajectory without Contact

Intuitively, we need to model the probability that the robot executes the trajectory τ without touching the object in a given pose $\mathbf{x}$. The Boolean function $\xi(\mathbf{x}, \tau)$ is used to indicate whether $\psi(\tau)$ intersects the volume of the object, Equation (5):

$$\xi(\mathbf{x}, \tau) = \begin{cases} 1, \text{ if } \psi(\tau) \text{ intersects the object in pose } \mathbf{x} \\ 0, \text{ otherwise} \end{cases} \tag{5}$$

Building upon [8], the contact-detection measurement is modeled as in Equation (6). α and β represent the false-negative and false-positive error rates for the contact detection.

$$P(z_{CD}|\mathbf{x}, \tau) = \begin{cases} P(z_{CD} = NC|\xi(\mathbf{x}, \tau) = 1) = \alpha \\ P(z_{CD} = C|\xi(\mathbf{x}, \tau) = 1) = 1 - \alpha \\ P(z_{CD} = C|\xi(\mathbf{x}, \tau) = 0) = \beta \\ P(z_{CD} = NC|\xi(\mathbf{x}, \tau) = 0) = 1 - \beta. \end{cases} \tag{6}$$

The false-negative error rate α corresponds to the probability of not measuring contact when the sweep intersects the volume occupied by the object. This coefficient captures two main sources of uncertainty. Firstly, when the robot measures $z_{CD} = NC$, it might actually be in contact with the object, but in some configuration that prevents the force-torque signals from reaching the thresholds. Figure 5a1 illustrates an example of such a false-negative measurement situation when the limited stiffness of the controller and the low surface friction allow the robot to contact the object and slide over it without exceeding the thresholds. Secondly, α also captures the uncertainty related to the mismatch between the object model and the actual object, which may cause the robot to intersect the polygonal mesh without measuring any real contact. An example of the latter situation is depicted in Figure 5a2. In general, α is a function of the intersection between $\psi(\tau)$ and the volume occupied by the object in pose $\mathbf{x}$. In this work, this computation is simplified assuming α as a function of the penetration length l, which is the intersection of the nominal trajectory τ and the volume of the object.

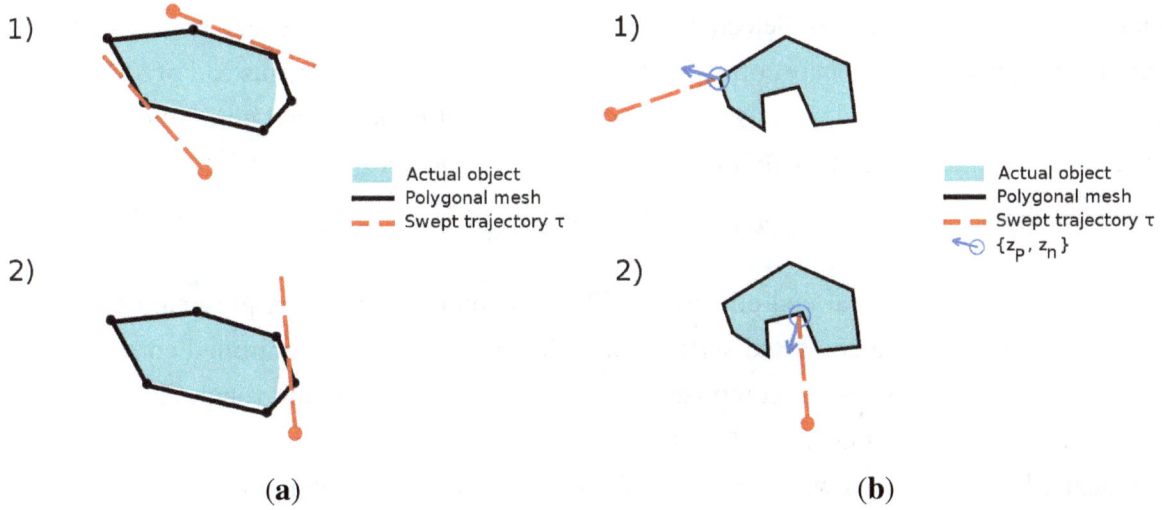

Figure 5. False-negative contact situations (**a**); and biased contact measurement situations (**b**). (a) Possible false-negative contact-detection situation: (1) the end effector touches the object sliding over it without exceeding the force threshold; (2) the end effector sweeps through the geometric model without touching the real object. (b) Examples of the contact state possibly leading to an inaccurate measurement: the end effector touches a vertex (Case 1) or a corner (Case 2). The measured normal does not correspond to any of the adjacent faces of on the object.

The false-positive error rate β encodes the probability of measuring contact when the sweep does not intersect the object volume. The threshold to detect contact is set on the resultant force and is parametrized with respect to the maximum standard deviation error among those of the force signals. This guarantees a lower false contact error rate than other thresholding policies, e.g., ellipsoidal or parallelepiped. In addition, we make the assumption of modeling the object with a polygonal mesh that is external or at most coincident to the nominal geometry. Moreover, the limited speed of the robot makes inertial effects minimal. Under these conditions, the probability of measuring a false contact is negligible. This corresponds to assuming $\beta \approx 0$.

In practice, the likelihood function P_{free} corresponding to a sweep without contact along τ and object pose $\mathbf{x}$ is expressed in Equation (7).

$$P_{free}(\mathbf{x}, \tau) = P(NC|\xi(\mathbf{x}, \tau)) \tag{7}$$

6.3. The Robot Makes Contact with the Object

Let us name τ_c the trajectory executed by the robot following τ up to pose ϕ_c on which the contact is detected. The measurement model expresses the probability of observing no-contact along τ_c, plus a contact in ϕ_c with touch measurement z_{touch}:

$$P_{contact}(\mathbf{x}, \tau) = P_{free}(\mathbf{x}, \tau_c)P(z_{CD}, z_{touch}|\mathbf{x}, \phi_c) \tag{8}$$

$P_{free}(\mathbf{x}, \tau_c)$ can be calculated as in Equation (7). By the definition of conditional probability:

$$P(z_{CD}, z_{touch}|\mathbf{x}, \phi_c) = P(z_{touch}|\mathbf{x}, \phi_c, z_{CD})P(z_{CD}|\mathbf{x}, \phi_c) \tag{9}$$

where $P(z_{CD}|\mathbf{x}, \phi_c)$ is calculated as in Equation (6). Formally, ϕ_c corresponds to the single-frame trajectory at which a contact is detected. However, in case of non-perfect proprio-perception or measurement capabilities, this may correspond to a portion of the trajectory instead of a single frame.

The touch measurement z_{touch} is comprised of both contact point z_P and normal vector z_n. This is observed only when the robot actually touches the object, *i.e.*, when $z_{CD} = C$:

$$P(z_{touch}|\mathbf{x}, \phi_c, z_{CD}) = P(z_P, z_n|\mathbf{x}, \phi_c, z_{CD} = C) \tag{10}$$

In the case of non-spherical end-effector, in [20], techniques have been presented to estimate the geometric parameters of the contacted surface through continuous force-controlled compliant motions. Since our final application case is comprised of harsh scenarios presenting rough and possibly soiled surfaces, we may not assume the robot to be capable of performing compliant motions in such environments. Instead, actions with the spherical pin used as a depth-probing finger are here considered as motion primitives. Unfortunately, with such actions, it is possible to experience contact states leading to inaccurate measurements, e.g., when the pin touches the object over a vertex, an edge or a corner. In this case, the normal vector measurement is significantly biased with respect to the nominal vector of the adjacent faces, as depicted in Figure 5b.

As in [4,8,10], we adopt the simplifying approximation of supposing both measurements z_P and z_n independent if conditioned on $\mathbf{x}, \phi_c$ and z_{CD}. Therefore, the joint measurement probability can be expressed as in Equation (11).

$$P(z_P, z_n|\mathbf{x}, \phi_c, z_{CD}) = P(z_P|\mathbf{x}, \phi_c, z_{CD})P(z_n|\mathbf{x}, \phi_c, z_{CD}) \tag{11}$$

With the object modeled as a polygonal mesh $\{f_i, n_i\}$ in pose $\mathbf{x}$ and with the robot in pose ϕ_c, one can express the likelihood functions encoding the probability of each patch $< f_i, n_i >$ to cause z_P and z_n as in Equations (12) and (13). Specifically, the uncertainty on z_{touch} is due to the noise acting on the six channels of the force-torque signal, to vertex or edge contact interaction (as depicted in Figure 5b) and surface imperfections. Such noise is assumed to be Gaussian for both z_P and z_n.

$$P(z_P|f_i, \phi_c, C) = \frac{1}{\sqrt{2\pi}\sigma_p} \exp\left[-\frac{1}{2}\frac{dist(z_P, f_i)^2}{\sigma_p^2}\right] \tag{12}$$

and:

$$P(z_n|n_i, \phi_c, C) = \frac{1}{\sqrt{2\pi}\sigma_n} \exp\left[-\frac{1}{2}\frac{||z_n - n_i||^2}{\sigma_n^2}\right] \tag{13}$$

The operator $dist(z_P, f_i)$ returns the distance between a measured point z_P and a face f_i of the polygonal mesh, whereas $||z_n - n_i||$ is the norm of the difference between the measured normal vector and the patch normal vector. After experimental tests, the Gaussian assumption for Equations (12) and (13) proves legitimate to model the uncertainty due to the measurement system and the non-perfect surface of the object. Yet, this model does not effectively capture the edge and vertex contact states depicted in Figure 5b. This limitation results in some estimation outliers, as further explained in Section 7.1.

The likelihood functions in Equations (12) and (13) are written with respect to a single $< f_i, n_i >$ patch, while the object model is composed by a whole mesh. Since it is necessary to express the likelihood functions with respect to the whole mesh in pose $\mathbf{x}$, a maximum-likelihood approach can then be applied: the patch $< f^*, n^* >$ that maximizes the product of both the contact and normal likelihoods is the one used to calculate the likelihood, as in Equation (14).

$$< f^*, n^* >= \arg \max_{<f_i, n_i>} (P(z_P | f_i, \phi_c, C) P(z_n | n_i, \phi_c, C)) \qquad (14)$$

This association strategy is also referred to as hard assign [21], and it implies an over-estimation of the likelihood function for the pose $\mathbf{x}$:

$$P(z_{touch} | \mathbf{x}, \phi_c) = P(z_P | f^*, \phi_c, C) P(z_n | n^*, \phi_c, C) \qquad (15)$$

Alternatively, one may also follow a soft assign approach, considering the likelihood of the pose as a linear combination of the likelihood of its faces. Since computing Equations (12) and (13) prove to be computationally expensive, the latter option is neither adopted in the literature nor in this work.

Nevertheless, even using Equation (14), computing contact and normal probabilities may become cumbersome when using large meshes. Instead, as already proposed by [4], one may select the face that only maximizes $P(z_P | f_i, \phi_c, C)$, as in Equation (16).

$$f^* = \arg \max_{f_i} (P(z_P | f_i, \phi_c, C)) \qquad (16)$$

This allows the estimator to speed up the calculation, since Equation (13) is computed only once, but it prevents the model from considering the normal likelihood when selecting the face. However, this simplification may be considered legitimate when $\sigma_p \ll \sigma_n$ (for instance, with a difference of orders of magnitude). For the sake of clarity, Figure 6 illustrates an example of the object mesh with a measured contact and normal vector affected by Gaussian noise. Being the closest to the contact point, face f_2 is selected to represent the pose of the object in terms of the likelihood function Equation (17):

$$P(z_{touch} | \mathbf{x}, \phi_c, C) = P(z_P | f_2, \phi_c, C) P(z_n | n_2, \phi_c, C). \qquad (17)$$

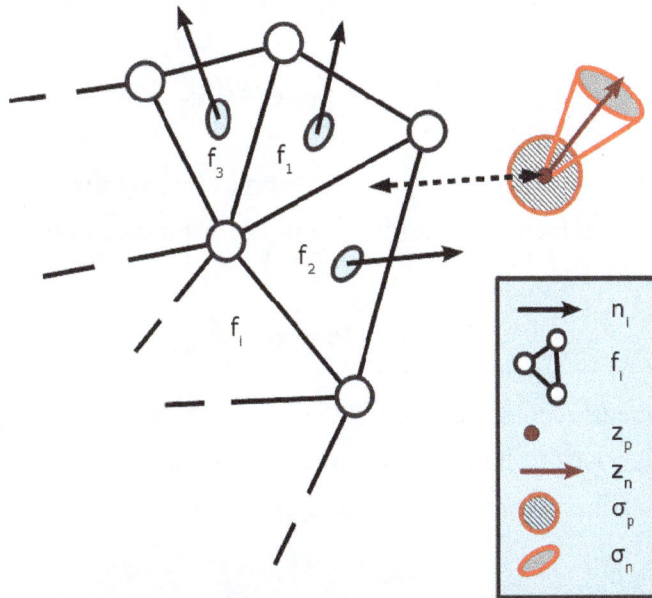

Figure 6. Polygonal mesh representing the object as a set of face-normal tuples $< f_i, n_i >$. The contact and normal vector are modeled as affected by Gaussian noise, represented by a sphere of radius σ_p and a cone of radius σ_n, respectively.

6.4. Inference Scheme

The state $\mathbf{x}_t \in X$ represents the pose of the object with respect to the robot frame of reference at time t. Its evolution over time is described as a hidden Markov model, *i.e.*, a dynamical stochastic system where $\mathbf{x}_t$ depends only on the previous state $\mathbf{x}_{t-1}$. This derives from the Markov assumption of state completeness, *i.e.*, no variable prior to $\mathbf{x}_{t-1}$ may influence the evolution of the future states. The state-measurement model can then be summarized as:

$$\mathbf{x}_t = g(\mathbf{x}_{t-1}) + \delta_t \tag{18}$$

$$z_t = m(\mathbf{x}_t) + \epsilon_t \tag{19}$$

where $g(.)$ and $m(.)$ are the transition model and measurement functions and δ_t and ϵ_t are transition and measurement noises. In a Bayesian context, the state is inferred updating a posterior distribution over the configuration space accounting for the measurement z_t, which is modeled as a set of independent random variables z_t drawn from the conditional probability distribution $P(z_t|\mathbf{x}_t)$, also referred to as likelihood. The motion model is encoded by the transition probability $P(\mathbf{x}_t|\mathbf{x}_{t-1})$. Our estimation objective is to infer the posterior probability over the state given the available measurement $P(\mathbf{x}_t|z_{0:t})$. In this work, the notation proposed in [22] is taken as a reference to represent the state transition and update. Specifically, $\bar{bel}(\mathbf{x}_t)$ and $bel(\mathbf{x}_t)$ are used to denote these two conditional probabilities, as in Equations (20) and (21).

$$\bar{bel}(x_t) = P(\mathbf{x}_t|z_{0:t-1}) \tag{20}$$

$$bel(x_t) = P(\mathbf{x}_t|z_{0:t}) \tag{21}$$

A Bayesian filter requires two steps to build the posterior probability, a prediction and an update. During prediction, the belief is calculated as in Equation (22), projecting the state from time $t-1$ to time t using the transition model.

$$\bar{bel}(x_t) = \int_{\mathbf{x}_{t-1}} P(\mathbf{x}_t|\mathbf{x}_{t-1})\bar{bel}(\mathbf{x}_{t-1})d\mathbf{x}_{t-1} \tag{22}$$

During the update, the posterior probability is updated multiplying the belief $\bar{bel}$ by the measurement likelihood and the normalization factor η. Intuitively, this corresponds to a correction of the model-based prediction based on measured data:

$$bel(x_t) = \eta P(z_t|\mathbf{x}_t)\bar{bel}(\mathbf{x}_t) \tag{23}$$

In our application, the measurement likelihood $P(z_t|\mathbf{x}_t)$ is calculated as in Equations (7) and (8). In the case of motion without contact, the transition model is the identity matrix. In the case of contact, the transition model is a diagonal covariance matrix $\Sigma_{contact}$:

$$\Sigma_{contact} = \sigma_{contact}^2 I \tag{24}$$

where $\sigma_{contact}$ is used to model the uncertainty on the object pose introduced by the contact. The normalization factor η is calculated as:

$$\eta = \frac{1}{\int_{\mathbf{x}_t} P(z_t|\mathbf{x}_t)\bar{bel}(\mathbf{x}_t)d\mathbf{x}_t} \tag{25}$$

and assures that the posterior probability sums up to one. Since we may not assume the posterior distribution over the state to be unimodal and the measurement is not linear, Equations (12) and (13), a particle filter algorithm is adopted. Its implementation is presented in pseudocode in Algorithm 2, with χ representing the particle set. χ_0 is initialized with samples drawn uniformly over the configuration space. Every time an action is completed, the algorithm is run. In the case of contact, the weights are calculated according to Equation (8). In the case of the trajectory finishing without contact, the weight is updated according to Equation (7).

Algorithm 2 Object localization with particle filter.

1: $\chi_t = [\,]$
2: **for** $i = 1 : N$ **do**
3:      sample $x_t^{[i]} \sim P(x_t | x_{t-1}^{[i]})$ ▷ Draw particle from transition model
4:      **if** contact **then**   ▷ update contact weight
5:          $w_t^{[i]} = P_{free}(x_t^{[i]}, \tau_c) P(C, z_{touch} | x_t^{[i]}, \phi_c)$
6:      **else**   ▷ Update no-contact weight
7:          $w_t^{[i]} = P_{free}(x_t^{[i]}, \tau)$
8:      **end if**
9:      add $x_t^{[i]}$ to $\bar{\chi}_t$
10: **end for**
11: **for** $i = 1 : N$ **do**
12:      draw index m from $\{x_t^{[i]}, w_t^{[i]}\}$ ▷ Resampling with replacement
13:      add $x_t^{[m]}$ to χ_t
14: **end for**
15: **for** $i = 1 : N$ **do**
16:      $\tilde{w}_t^{[i]} = \dfrac{w_t^{[i]}}{\sum_0^N w_t^{[i]}}$ ▷ Normalization
17: **end for**

---

7. Experimental Results

Experimental results are presented in the context of touch-based active sensing using the robot setup described in the previous section. Focusing on industrial relevance, this work improves the state-of-the-art:

- localizing objects up to industrial geometric complexity;
- localizing an object with high initial uncertainty, keeping the estimation problem tractable using the DOF decoupling task graph model;
- speeding up an object localization task using the `act-reason` algorithm setting the allocated time as a function of the current uncertainty.

7.1. Localization of Objects of Increasing Complexity

To the best of our knowledge, literature examples presented case studies with polygonal meshes up to about 100 faces [8,10]. Here, we contribute to the state-of-the-art localizing objects up to industrial complexity. The application is specified with a task graph with as a single task:

- configuration space: the pose of the object $\mathbf{x} = \{x, y, \theta_z\}$ defined with respect to the robot frame of reference. Initial uncertainty: (0.1 m, 0.1 m, 1 rad).
- Information space: $P(\mathbf{x}|z)$.
- Object model: polygonal mesh.
- Action space: user-defined actions aiming at touching the largest faces of the object in the best-estimate pose. Sensing: contact detection and touch measurement.
- Inference scheme: particle filter.
- Action-selection scheme: random selection.

Table 2 presents the objects used in the experiment: the solid rectangle (obj$_{SR}$), the v-block (obj$_{VB}$), the cylinder-box (obj$_{CB}$) and the ergonomy test mock-up (obj$_{ET}$). Together with their properties, mesh and action space, the table shows the obtained localization results. In this test, $\sigma_P = 0.005$ m and $\sigma_n = 0.1$.

The following protocol was followed for each run:

1. Actions were repeated until 80% of the probability mass on the target position (indicated in red in Table 2) was concentrated around a 5-mm radius, which is consistent with the tolerance allowed in typical robot operations, e.g., for grasping and manipulation applications. The probability threshold was chosen to be the same as in [4].
2. Once the 80% confidence threshold was reached, a peg-in-hole sequence aiming at the target on the object was performed recording the final position of the spherical end-effector.
3. The end-effector was manually positioned on the target spot and the end-effector position recorded.

The considered ground truth about the position of the target was biased by the uncertainty related to human vision. After empirical observation, this can be estimated on the order of 2 mm. Even if biased, this may be considered as a genuine mean to evaluate the localization accuracy. Table 2 presents the localization results in terms of final $\{x, y\}$ target position error with respect to the observed ground truth (Loc.error). The mean $\bar{e}$ and the standard deviation of the error norm σ_e are also reported.

Overall, the tests carried out on the obj$_{VB}$ present the best accuracy in terms of mean error norm, even though an outlier was recorded due to a vertex contact, as shown in Figure 5a. Unsurprisingly, tests on obj$_{ET}$ present the highest mean and standard deviation error, yet with most of the bias concentrated along the x axis. This is likely due to a non-negligible misalignment between the polygonal mesh and the actual geometry of the object, which experienced plastic displacement due to heavy usage previous to this test.

Table 2. Summary of the object localization tests. For each studied object, the table reports geometric and dynamic characteristics, action space, localization error, mean error norm $\bar{e}$ and error norm standard deviation σ_e.

Object	Obj_{SR}	Obj_{VB}	Obj_{CB}	Obj_{ET}
Size (m)	$0.15 \times 0.15 \times 0.1$	$0.175 \times 0.15 \times 0.05$	$0.35 \times 0.125 \times 0.08$	$0.44 \times 0.14 \times 0.14$
Vertices	8	96	191	915
Faces	12	188	376	1,877
Weight (kg)	1.2	3	5.4	10.5
Model				
Actions				
Loc. error				
$\bar{e}$ **(m)**	0.005	0.0043	0.0053	0.0073
σ_e **(m)**	0.0025	0.0034	0.0052	0.0048

7.2. Touch-Based Localization with High Initial Uncertainty

Localising an object in 3DOFs over a table with a tolerance of 5 mm and initial uncertainty (0.4 m, 0.4 m, 2π) may become computationally expensive and prevent online action-selection. Prior work examples of similar tasks [4,8,12,14] suffered from this curse of dimensionality, requiring several minutes to perform the task. In this section, we show how the complexity of the task can be reduced adopting an empirical task graph derived from the human experiment presented in [13] and applying it to the localization of obj_{VB}. Specifically, the application is divided into two subtasks, S1 and S2, as detailed below and illustrated in Figure 7.

- Subtask S1:

  - configuration space C1: $\{x, y\}$ position of the object with respect to the robot.
  - information space: P($\mathbf{x}$|z).
  - object model: polygonal mesh of the bounding cylinder of obj_{VB}.
  - action space: sweep with contact detection.

– inference scheme: particle filter.

– action-selection scheme: zig-zag motions parametrized with respect to the table dimensions.

- Subtask S2:

  – configuration space C2: $\{x, y, \theta_z\}$ pose of the object with respect to the robot.

  – information space: posterior distribution over the object configuration space given the contact measurement.

  – object model: nominal polygonal mesh of $\mathrm{obj_{VB}}$.

  – action space: sweeps with contact detection and touch measurements.

  – inference scheme:particle filter.

  – action-selection scheme: random actions aiming at the four faces of the object in its best-estimate configuration.

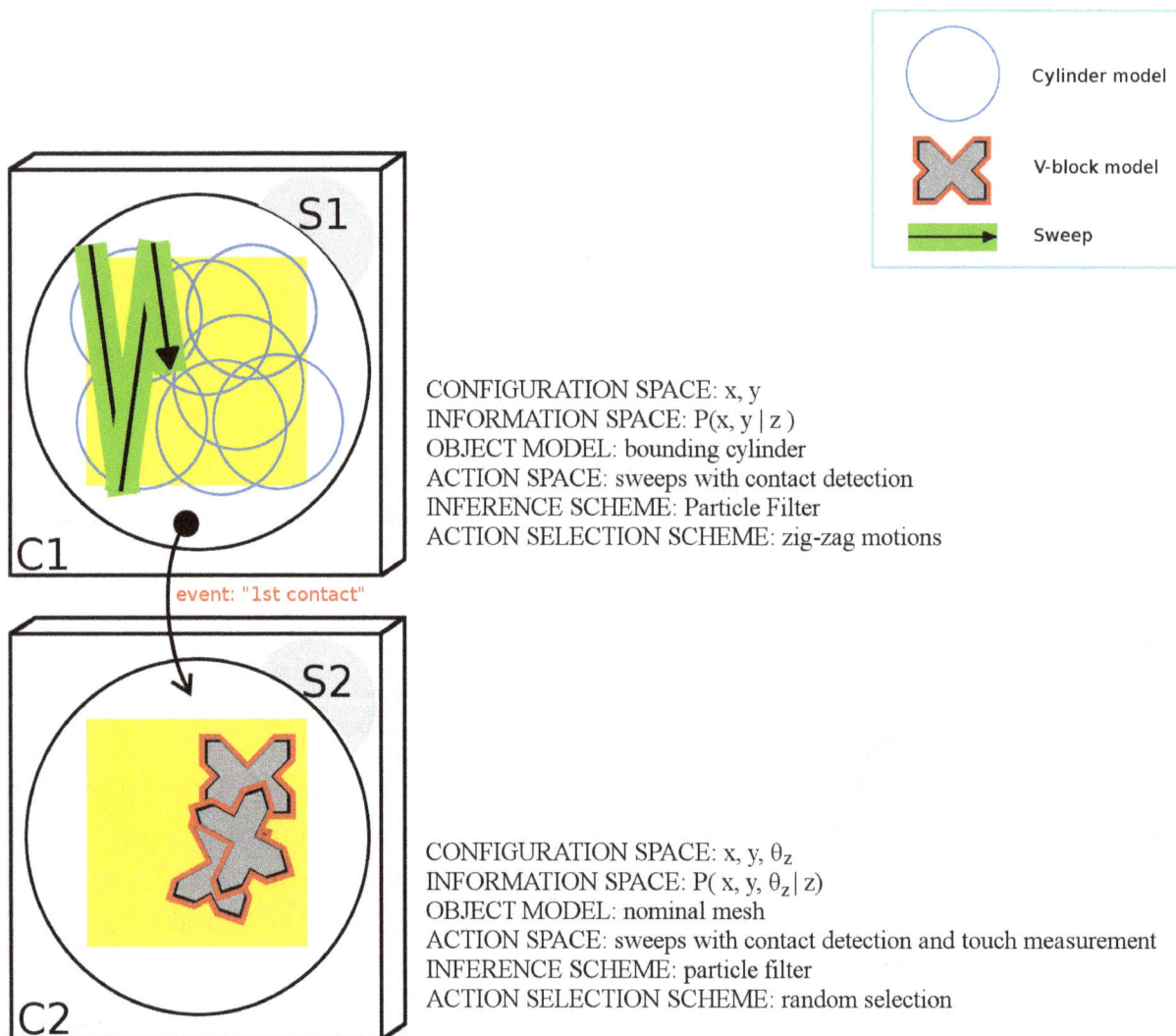

CONFIGURATION SPACE: x, y
INFORMATION SPACE: P(x, y | z)
OBJECT MODEL: bounding cylinder
ACTION SPACE: sweeps with contact detection
INFERENCE SCHEME: Particle Filter
ACTION SELECTION SCHEME: zig-zag motions

CONFIGURATION SPACE: x, y, θ_z
INFORMATION SPACE: P(x, y, θ_z| z)
OBJECT MODEL: nominal mesh
ACTION SPACE: sweeps with contact detection and touch measurement
INFERENCE SCHEME: particle filter
ACTION SELECTION SCHEME: random selection

Figure 7. Task graph derived from the human experiment in [13]. The application is divided into an initial 2DOF subtask (S1) defined over configuration space C1 and a second subtask (S2) defined over configuration space C2.

S1 is aimed at reducing the uncertainty on x and y applying a zig-zag action-selection strategy that spans the table surface in a breadth-first fashion. The object is represented as the cylinder that bounds obj_{VB} and has radius ρ. Inference is performed through a particle filter. The action space is comprised of sweeps with contact detection. S2 is aimed at finely estimating the pose of the object, which is represented with its nominal mesh. Inference is performed with a particle filter. The action space is comprised of sweeps with contact detection and touch measurement aiming at 12 different faces of obj_{VB} (see Table 2). To capture geometric imperfections, α was set to 5 mm. The same protocol presented in Subsection 7.1 was followed. Fifteen trials were carried out with the object located in a random pose on top of the table. In all cases, task execution took less than 120 s, and the final error position of the end-effector is reported in Figure 8.

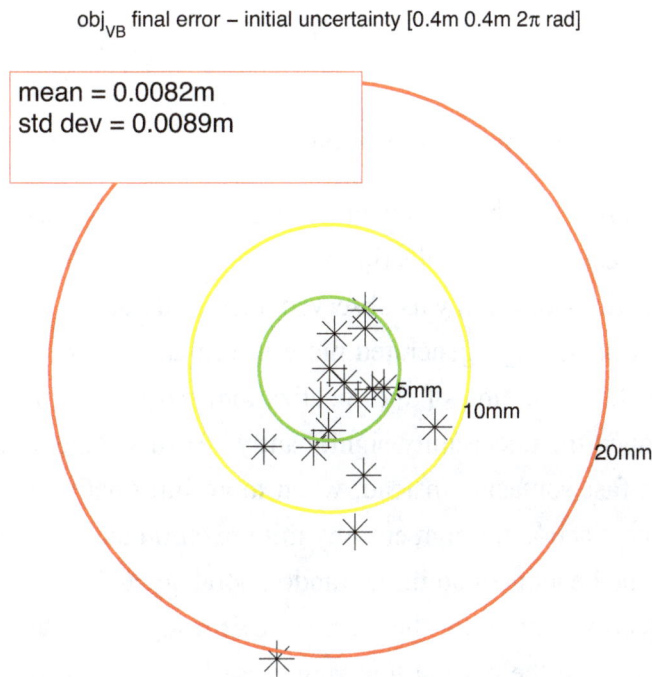

obj_{VB} final error – initial uncertainty [0.4m 0.4m 2π rad]

mean = 0.0082m
std dev = 0.0089m

Figure 8. Peg-in-hole error during v-block localization with DOF decoupling with (0.4 m, 0.4 m, 2π) initial uncertainty.

7.3. Reducing Task Execution Time with the `act-reason` Algorithm

In the context of touch-based localization, different action-selection schemes have been adopted in the literature, including random selection [12], entropy minimization [4] and Kullback–Leibler divergence [8]. However, in these examples, actions were selected from a predefined action space, with a significant impact of computation over task execution time. Here, the focus is on time efficiency, and we present a concrete example of how the `act-reason` algorithm can be adopted to improve it with respect to random selection and entropy minimization. According to Equation (2), the reward corresponds to the difference of the entropy H of the prior distribution at time t and the entropy of the posterior distribution after the action a is executed:

$$r(a) = H(P(\mathbf{x}_t)) - H(P(\mathbf{x}_{t+1})|a). \tag{26}$$

In this work, the particle filter formulation of entropy proposed by [5] is adopted.

Formally, the studied localization application of obj_{VB} is represented by the six following primitives of the active sensing task model.

- Configuration space: the pose of the object $\mathbf{x} = \{x, y, \theta_z\}$ defined with respect to the robot frame of reference. Initial uncertainty: (0.1 m 0.1 m 1 rad).
- Information space: $P(\mathbf{x}|z)$.
- Object model: polygonal mesh of obj_{VB}.
- Action space: user-defined actions aiming at touching the largest faces of the object in its best-estimate pose. Sensing: contact detection and touch measurements.
- Inference scheme: particle filter.
- Action-selection scheme:

  1. random selection
  2. entropy minimization
  3. `act-reason` with entropy minimization.

In this context, task progress can be evaluated through the current uncertainty on the pose of the object. During preliminary tests, random selection was compared to entropy minimization. The better performance of a random action-selection was observed, especially at the beginning of the localization, while an entropy minimization strategy generated more informative actions, yet taking longer for their evaluation. Intuitively, at the beginning of the localization, very little information is available, and any contact is likely to reduce the uncertainty significantly. At this stage, one would wish to follow a random selection to obtain fast contacts. Instead, when more information is available, spending more time reasoning to find the best action through entropy minimization tends to pay off.

In order to exploit both the benefits of an initial random strategy and the informative actions obtained with an entropy-based selection, we adopt the `act-reason` algorithm to vary the allocated time to reason and execute depending on the current uncertainty, as in Equation (27). Specifically, we set the allocated time t_{alloc} to vary between the average time required by a random action $\hat{t}_{rand}$ and that of a full-resolution entropy minimization $\hat{t}_{ent}$, considering both computation and execution. For the sake of simplicity, this variation is set linear in the progress metric Π:

$$t_{alloc} = \hat{t}_{rand} + \Pi \left(\hat{t}_{ent} - \hat{t}_{rand} \right). \tag{27}$$

In our implementation, Π is defined as the ratio between the current probability mass within the 5-mm tolerance and the desired value to call the localization done.

$$\Pi = \frac{P(\mathbf{x} \in tol)}{P(\mathbf{x} \in tol)_{final}}. \tag{28}$$

Intuitively, this corresponds to adjusting the robot behavior from fully random when the uncertainty is high, to fully info-gathering when the uncertainty is low and more informative actions improve convergence.

Figure 9 presents the results in terms of information gain *vs.* time during a v-block localization with initial uncertainty of (0.1 m, 0.1 m, 1 rad). Applying the `act-reason` algorithm to vary t_{alloc} as in

Equation (27), a faster second contact and a lower execution time is recorded. In Table 3, the average task-execution time $\bar{t}_{tot}$ and the average time to second contact $\bar{t}_{2nd}$ are reported. With act-reason, task-execution time is reduced by 52% with respect to entropy minimization and by 35% with respect to random selection.

Table 3. V-block localization with initial uncertainty (0.1 m, 0.1 m, 1 rad): comparison of task-execution time and time to second contact with act-reason, entropy minimization and random selection.

	Act-Reason	H Min.	Random
$\bar{t}_{tot}$ (s)	40.1	61.0	54.0
$\bar{t}_{2nd}$ (s)	8.7	12.0	8.4

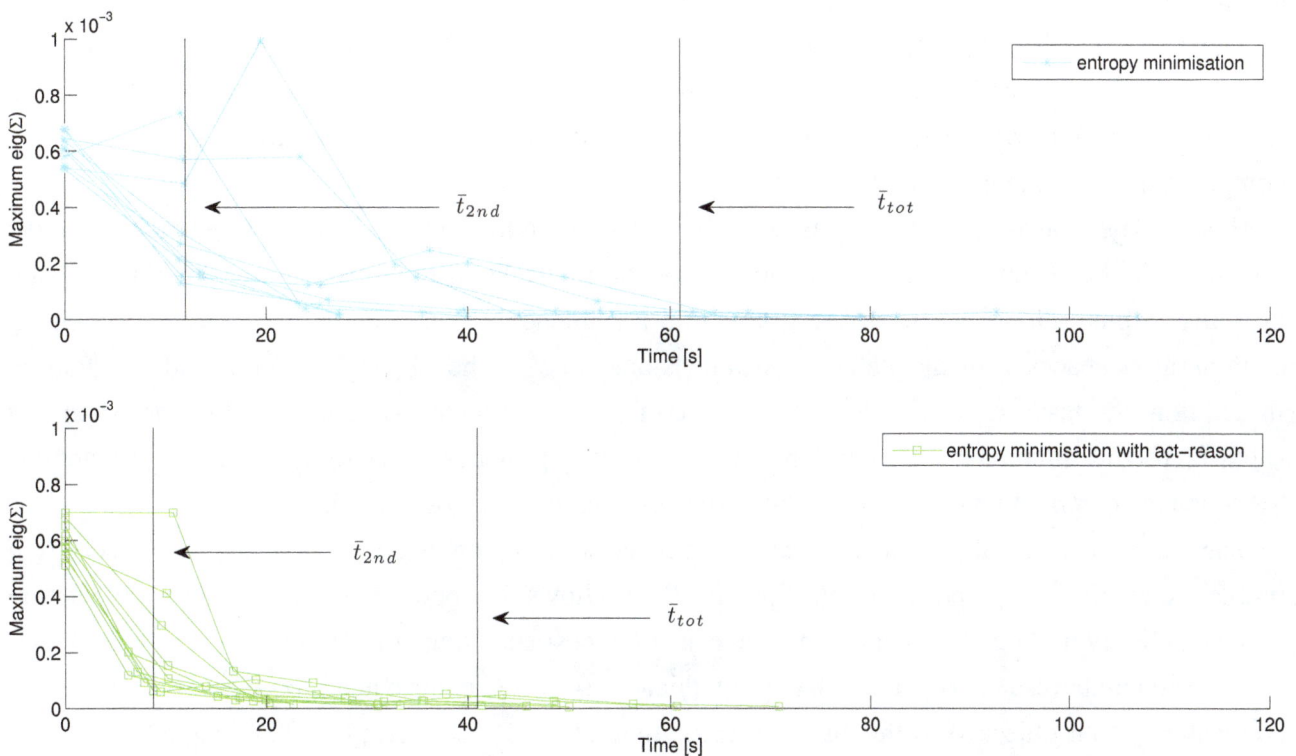

Figure 9. V-block localization with initial uncertainty (0.1 m, 0.1 m, 1 rad). Progress *vs.* time with fixed action space (cyan) and variable action space using the act-reason algorithm (green).

8. Conclusions

Active sensing has been almost exclusively an academic research interest, with strong emphasis on search algorithms, with proofs of concept demonstrations on rather simple objects and without paying much attention to the overall time to realize the task. However, with sensor-based robots becoming more commonplace in industrial practice, there is a drive towards more performant active sensing task selection, on objects with the geometric complexity of industrial applications. Hence, there is a large demand to reduce the time it takes to compute, execute and process the active sensing actions.

The contribution of this paper that is expected to have the biggest impact on such a time reduction is the insight that the state-of-the-art until now has been using the implicit assumption that an active sensing task that needs to localize geometric features on objects in the six-dimensional configuration space of those features' position and orientation should also do all of its computations in that same six-dimensional configuration space. However, our extensive observations of humans performing active sensing showed that they are very good at adapting the dimension of the search space to the amount of information that has already been found. The main gain in speed then comes from using only two- or three-dimensional motions to find the first contact with the object; after that, similar low-dimensional searches can be used, for example to scan a planar face of the object for the feature. Many different strategies are possible for how to choose such lower-dimensional searches, but the major observation is that humans only switch to the eventual six-dimensional search space towards the very end of the active sensing task.

The other contributions of this paper also have an impact on the desired reduction in the computation of active sensing actions, by suggesting a formal model of active sensing, in which all relevant aspects are represented separately, as well as the relationships between these aspects that a particular active sensing computation must take into account.

First, and generalizing from previous examples in the literature, a model-level representation of active sensing tasks has been presented. The novelty of the model is in the introduction of the task graph primitive, which allows one to divide a task into a sequence of subtasks defined over less complex configuration spaces than the default six-dimensional one of the objects' position and orientation. In addition, the task graph also stores the choice of search strategies to be used in the corresponding subtask. The information that becomes available in this way provides online reasoners with more options to reduce the computational complexity, thus speeding up the localization task.

The `act-reason` algorithm has been formulated and implemented to solve the action-selection problem once the task graph is chosen. Specifically, it allows the robot to reason over a variable action space, explicitly trading off information gain, execution cost and computation time.

Experimental results on touch-based applications have contributed to the state-of-the-art: (i) localizing four objects of different complexity up to about 10^3 vertices; (ii) localizing a solid object with high initial uncertainty without suffering from the curse of dimensionality thanks to the task graph structure; and (iii) improving the time efficiency of object localization using `act-reason`, setting the allocated time as a function of the current uncertainty.

9. Discussion and Future Work

Applying an empirical task graph observed during a human experiment, significant complexity reduction was achieved on a 3DOF force-based localization application. Enhancing these results by defining the task graph out of a constrained optimization would represent a remarkable extension of our findings. In this regard, the distributional clauses particle filter framework [23] might be adopted to encode topological relations between different objects, such as {inside, on top, beside}, in order to define the structure of the task graph from the CAD model of the environment.

The object model presented in Section 3.1 takes into account geometric properties. This information is sufficient to perform active sensing with probes, such as vision or force sensors. However, this model could be extended to take into account additional information, such as dynamic or thermic properties of the object, which can be exploited by other types of sensors.

An action space composed of more complex touch primitives (e.g., compliant motion for contour following) could exploit the edges and corners of the object to gather more information for the estimation.

Acknowledgments

This work was sponsored by CEA, List, Interactive Robotics Laboratory. H.B. acknowledges the support from the University of Leuven Geconcerteerde Onderzoeks-Actie, Global real-time optimal control of autonomous robots and mechatronic systems, and from the European Union's 7th Framework Programme (FP7/2007–2013) projects ROSETTA(FP7-230902, Robot control for skilled execution of tasks in natural interaction with humans; based on autonomy, cumulative knowledge and learning), and RoboHow.Cog (FP7-288533, Web-enabled and experience-based cognitive robots that learn complex everyday manipulation tasks).

Author Contributions

The work reported here took place in the context of the PhD project of Niccoló Tosi, who is credited with the vast majority of the work. Olivier David defined the context that drove this research and verified its relevance for industrial applications. Herman Bruyninckx contributed to the model formalization of active sensing tasks. All authors discussed the results and commented on the manuscript at all stages.

Conflicts of Interest

The authors declare no conflict of interest.

References

1. Geffard, F.; Garrec, P.; Piolain, G.; Brudieu, M.A.; Thro, J.F.; Coudray, A.; Lelann, E. TAO2000 V2 computer-assisted force feedback telemanipulators used as maintenance and production tools at the AREVA NC-La Hague fuel recycling plant. *J. Field Robot.* **2012**, *29*, 161–174.
2. David, O.; Russotto, F.X.; Simoes, M.D.S.; Measson, Y. Collision avoidance, virtual guides and advanced supervisory control teleoperation techniques for high-tech construction: Framework design. *Autom. Constr.* **2014**, *44*, 63–72.
3. De Geeter, J.; van Brussel, H.; de Schutter, J.; Decreton, M. Recognising and locating objects with local sensors. In Proceedings of the 1996 IEEE International Conference on Robotics and Automation, Minneapolis, MN, USA, 22–28 April 1996; Volume 4, pp. 3478–3483.
4. Hsiao, K. Relatively Robust Grasping. Ph.D. Thesis, Massachusetts Institute of Technology, Department of Electrical Engineering and Computer Science, Cambridge, MA, USA, 2009.

5. Nikandrova, E.; Laaksonen, J.; Kyrki, V. Towards informative sensor-based grasp planning. *Robot. Auton. Syst.* **2014**, *62*, 340–354.

6. Barragan, P.R.; Kaelbling, L.P.; Lozano-Perez, T. Interactive Bayesian identification of kinematic mechanisms. In Proceedings of the 2014 IEEE International Conference on Robotics and Automation (ICRA), Hong Kong, China, 2014; pp. 2013–2020.

7. Taguchi, Y.; Marks, T.; Hershey, J. Entropy-based motion selection for Touch-based registration using Rao-Blackwellized particle filtering. In Proceedings of the 2011 IEEE/RSJ International Conference on Intelligent Robots and Systems (IROS), San Francisco, CA, USA, 25–30 September 2011; pp. 4690–4697.

8. Hebert, P.; Burdick, J.; Howard, T.; Hudson, N.; Ma, J. Action inference: The next best touch. In Proceedings of the RSS 2012 Mobile Manipulation Workshop, Sydney, Australia, July 2012.

9. Gadeyne, K.; Bruyninckx, H. Markov Techniques for Object Localization With Force-Controlled Robots. In Proceedings of the ICAR, Budapest, Hungary, 22–25 August 2001.

10. Petrovskaya, A.; Khatib, O. Global Localization of Objects via Touch. *IEEE Trans. Robot.* **2011**, *27*, 569–585.

11. Gadeyne, K.; Lefebvre, T.; Bruyninckx, H. Bayesian Hybrid Model-State Estimation Applied To Simultaneous Contact Formation Detection and Geometrical parameter Estimation. *Int. J. Robot. Res.* **2005**, *24*, 615–630.

12. Petrovskaya, A.; Khatib, O.; Thrun, S.; Ng, A. Bayesian estimation for autonomous object manipulation based on tactile sensors. In Proceedings of the 2006 IEEE International Conference on Robotics and Automation, Orlando, FL, USA, 15–19 May 2006; pp. 707–714.

13. Tosi, N.; David, O.; Bruyninckx, H. DOF-Decoupled Active Force Sensing (D-DAFS): A human-inspired approach to touch-based localization tasks. In Proceedings of the 2013 International Conference on Advanced Robotics (ICAR), Montevideo, Uruguay, 25–29 November 2013; pp. 1–8.

14. Lefebvre, T.; Bruyninckx, H.; de Schutter, J. Task planning with active sensing for autonomous compliant motion. *Int. J. Robot. Res.* **2005**, *24*, 61.

15. Tosi, N.; David, O.; Bruyninckx, H. Action Selection for Touch-based Localization Trading off Information Gain and Execution Time. In Proceedings of the 2014 International Conference on Robotics and Automation, Hong Kong, China, 31 May–7 June 2014.

16. Krainin, M.; Curless, B.; Fox, D. Autonomous generation of complete 3D object models using next best view manipulation planning. In Proceedings of the 2011 IEEE International Conference on Robotics and Automation (ICRA), Shanghai, China, 9–13 May 2011; pp. 5031–5037.

17. Potthast, C.; Sukhatme, G.S. A probabilistic framework for next best view estimation in a cluttered environment. *J. Vis. Commun. Image Represent.* **2014**, *25*, 148–164.

18. Hsiao, K.; Kaelbling, L.P.; Lozano-Pérez, T. Robust grasping under object pose uncertainty. *Auton. Robot.* **2011**, *31*, 253–268.

19. Kitagaki, K.; Suehiro, T.; Ogasawara, T. Monitoring of a pseudo contact point for fine manipulation. In Proceedings of the 1996 IEEE/RSJ International Conference on Intelligent Robots and Systems' 96 (IROS 96), Osaka, Japan, 4–8 November 1996; Volume 2, pp. 757–762.

20. de Schutter, J.; Bruyninckx, H.; Dutré, S.; De Geeter, J.; Katupitiya, J.; Demey, S.; Lefebvre, T. Estimating first-order geometric parameters and monitoring contact transitions during force-controlled compliant motion. *Int. J. Robot. Res.* **1999**, *18*, 1161–1184.

21. Kearns, M.; Mansour, Y.; Ng, A.Y. An information-theoretic analysis of hard and soft assignment methods for clustering. In *Learning in Graphical Models*; Springer: Berlin, Germany, 1998; pp. 495–520.

22. Thrun, S.; Burgard, W.; Fox, D. *Probabilistic Robotics (Intelligent Robotics and Autonomous Agents)*; The MIT Press: Cambridge, MA, USA, 2005.

23. Nitti, D.; de Laet, T.; de Raedt, L. Distributional Clauses Particle Filter. In *Machine Learning and Knowledge Discovery in Databases*; Springer: Berlin, Germany, 2014; pp. 504–507.

6

A Computational Model of Human-Robot Spatial Interactions Based on a Qualitative Trajectory Calculus

Christian Dondrup [1,*], **Nicola Bellotto** [1], **Marc Hanheide** [1], **Kerstin Eder** [2] **and Ute Leonards** [3]

[1] School of Computer Science, University of Lincoln, Brayford Pool, LN6 7TS Lincoln, UK;
E-Mails: nbellotto@lincoln.ac.uk (N.B.); mhanheide@lincoln.ac.uk (M.H.)

[2] Department of Computer Science, University of Bristol, Merchant Venturers Building, Woodland Road, Clifton, BS8 1UB Bristol, UK; E-Mail: kerstin.eder@bristol.ac.uk

[3] School of Experimental Psychology, University of Bristol, 12A Priory Road, Clifton, BS8 1TU Bristol, UK; E-Mail: ute.leonards@bristol.ac.uk

* Author to whom correspondence should be addressed; E-Mail: cdondrup@lincoln.ac.uk

Academic Editor: Huosheng Hu

Abstract: In this paper we propose a probabilistic sequential model of Human-Robot Spatial Interaction (HRSI) using a well-established Qualitative Trajectory Calculus (QTC) to encode HRSI between a human and a mobile robot in a meaningful, tractable, and systematic manner. Our key contribution is to utilise QTC as a state descriptor and model HRSI as a probabilistic sequence of such states. Apart from the sole direction of movements of human and robot modelled by QTC, attributes of HRSI like proxemics and velocity profiles play vital roles for the modelling and generation of HRSI behaviour. In this paper, we particularly present how the concept of proxemics can be embedded in QTC to facilitate richer models. To facilitate reasoning on HRSI with qualitative representations, we show how we can combine the representational power of QTC with the concept of proxemics in a concise framework, enriching our probabilistic representation by implicitly modelling distances. We show the appropriateness of our sequential model of QTC by encoding different HRSI behaviours observed in two spatial interaction experiments. We classify these encounters, creating a comparative measurement, showing the representational capabilities of the model.

Keywords: qualitative trajectory calculus; human-robot spatial interaction; qualitative spatial relations; probabilistic sequential models; proxemics

1. Introduction

Currently used research and commercial robots are able to navigate safely through their environment, avoiding static and dynamic obstacles. However, a key aspect of mobile robots is the ability to navigate and manoeuvre safely around humans [1]. Mere obstacle avoidance is not sufficient in those situations because humans have special needs and requirements to feel safe and comfortable around robots. Human-Robot Spatial Interaction (HRSI) is the study of joint movement of robots and humans through space and the social signals governing these interactions. It is concerned with the investigation of models of the ways humans and robots manage their motions in vicinity to each other. These encounters might, for example, be so-called *pass-by* situations where human and robot aim to pass through a corridor trying to circumvent each other given spatial constraints. In order to resolve these kinds of situations and pass through the corridor, the human and the robot need to be aware of their mutual goals and have to have a way of negotiating who goes first or who goes to which side. Our work therefore aims to equip a mobile robot with understanding of such HRSI situations and enable it to act accordingly.

In early works on mobile robotics, humans have merely been regarded as static obstacles [2] that have to be avoided. More recently, the dynamic aspects of "human obstacles" have been taken into account, e.g., [3]. Currently, a large body of research is dedicated to answer the fundamental questions of HRSI and is producing navigation approaches which plan to explicitly move on more "socially acceptable and legible paths" [4–6]. The term "legible" here refers to the communicative–or interactive–aspects of movements which previously have widely been ignored in robotics research. Another specific requirement to motion planning involving more than one dynamic agent, apart from the sociability and legibility, is the incorporation of the other agent's intentions and movements into the robot's decision making. According to Ducourant *et al.* [7], who investigated human-human spatial behaviour, humans also have to consider the actions of others when planning their own. Hence, spatial movement is also about communication and coordination of movements between two agents–at least when moving in close vicinity to one another, e.g., entering each other's social or personal spaces [8].

From the above descriptions follow certain requirements for the analysis of HRSI that need to be fulfilled in order to equip a mobile robot with such an understanding of the interaction and the intention of its counterpart. Additionally, such a representation is used to evaluate and generate behaviour according to the experienced *comfort*, *naturalness*, and *sociability*, as defined by Kruse *et al.* [9], during the interaction. Hence, the requirements for such an optimal representation are:

Representing the qualitative character of motions including changes in direction, stopping or starting to move, *etc.* It is known that small movements used for prompting, e.g., [10], are essential for a robot to interpret the intention of the human and to react in a socially adequate way.

Representing the relevant attributes of HRSI situations in particular proxemics [8] (*i.e.*, the distance between the interacting agents), which we focus on in this paper. This is required to analyse the appropriateness of the interaction and to attribute intention of the implicitly interacting agents.

Ability to generalise over a number of individuals and situations. A robot requires this ability to utilise acquired knowledge from previous encounters of the same or similar type. A qualitative framework that is able to create such a general model, which still holds enough information to unambiguously describe different kinds of interactions but abstracts from metric space, facilitates learning and reasoning.

A tractable, concise, and theoretically well-founded model is necessary for the representation and underlying reasoning mechanisms in order to be deployed on an autonomous robot.

We have laid the first foundation for such an approach in a number of previous works investigating the suitability of applying a Qualitative Trajectory Calculus (QTC) to represent HRSI [11–14]. QTC is a formalism representing the relative motion of two points in space in a qualitative framework and offers a well-defined set of symbols and relations [15]. We are building on the results from [11,12] using a Markov model and hand crafted QTC state chains which has been picked up in [13,14] and evolved into a Hidden Markov Model (HMM) based representation of learned interactions. This paper offers a comprehensive overview of the QTC-based probabilistic sequential representation utilising the HMM, and focuses on its specific adoptions for the encoding of HRSI using real-world data. In this sense, we integrate our previous findings into a more unified view and evaluate the proposed model on a new and larger data set, investigating new types of interactions and compare these results to our previous experiment [13,14]. In particular, we assessed the generality of our model by not only testing it on a single robot type, but extended the set of experiments to include data from a more controlled study using a "mock-up" robot (later referred to as the "Bristol Experiment") in an otherwise similar setting. We argue that the proposed model is both rich enough to represent the selected spatial interactions from all our test scenarios, and that it is at the same time compact and tractable, lending itself to be employed in responsive reasoning on a mobile, autonomous robot.

As stated in our requirements, social distances are an essential factor in representing HRSI situations as indicated in Hall's proxemics theory [8] and numerous works on HRSI itself, e.g., [16]. However, QTC has been developed to represent the relative change in distance between two agents but it was never intended to model the absolute value. This missing representation deprives it of the ability to use proxemics to analyse the appropriateness of the interaction or to *generate* appropriate behaviour regarding HRSI requirements. To overcome this deficit and to highlight the interaction of the two agents in close vicinity to one another, we aim to model these distances using our HMM-based representation of QTC. Instead of modelling distance explicitly by expanding the QTC-state descriptors and including it as an absolute value, as e.g., suggested by Lichtenthäler *et al.* [17], we aim to model it implicitly and refrain from altering the used calculus to preserve its qualitative nature and the resulting generalisability, and simplicity. We utilise our HMM-based model and different variants of QTC to define transitions between a coarse and fine version of the calculus depending on the distance between human and robot. This not only allows to represent distance but also uses the richer variant of QTC only in close vicinity to the robot, creating a more compact representation and highlighting the interaction when both interactants are close enough to influence each other's movements. We are going beyond the use of hand crafted QTC state chains and a predefined threshold to switch between the different QTC variants as done in previous work [12], and investigate possible transitions and distances learned from real world data from two spatial interaction experiments. Therefore, one of the aims of this work is to investigate suitable transition states and distances or ranges of distances, comparing results from our two experiments, for our combined QTC model. We expect these distances to loosely correlate with Hall's personal space

($1.22\ m$) from observations made in previous work [14] which enables our representation to implicitly model this important social norm.

To summarise, the main contributions of this work are (i) a HMM-based probabilistic sequential representation of HRSI utilising QTC; (ii) the investigation of the possibility of incorporating distances like the crucial HRSI concept of proxemics [8] into this model; and (iii) enabling the learning of transitions in our combined QTC model and ranges of distances to trigger them, from real-world data. As a novel contribution in this paper we provide stronger evidence regarding the generalisability and appropriateness of the representation, demonstrated by using it to classify different encounters observed in motion-capture data obtained from different experiments, creating a comparative measurement for evaluation. Following our requirements mentioned above, we thereby aim to create a tractable and concise representation that is general enough to abstract from metric space but rich enough to unambiguously model the observed spatial interactions between human and robot.

2. Related Work

Qualitative spatial representations like QTC are used on a large scale in many different research areas and fields [18]. In our case, a probabilistic model of QTC state chains is used to describe interactions between a human and a robot in the spatial domain, *i.e.*, 2D navigation, which is why this section will focus on the different forms of representations used in HRSI and how they compare to the presented approach.

Representing spatial interaction is an important part of HRI in general and HRSI in particular where the vast majority of publications in the field of human-aware navigation represents interactions in metric space [9]. These representations are used mainly for *path planning* and *prediction* and employ the concept of *proxemics*, of which examples of currently used approaches will be shown in the following sections.

2.1. Proxemics

Before we go into detail on path planning and prediction we would like to introduce the concept of proxemics which is used in both experiments described in Section 5. We are adopting the definition of *personal space* and *social space* from [8]. In his work, E.T. Hall defines several distances and groups them into four different categories (in order of increasing distance between the interactants): *Intimate Space, Personal Space, Social Space,* and *Public Space*. Theses spaces are defined according to different factors, e.g., the ability to touch each other, loudness of voice while conversing, olfactory sensing, *etc.* Additionally, all of these spaces or distances are divided into two subgroups, *i.e.*, the *close phase* and the *far phase*. In the following, when speaking of personal space we refer to the area described by the close phase (1.22 m to 2.1 m) of the social space and the far phase of the personal space (0.76 m to 1.22 m). Previous work [14] has shown that this is the area providing the most promising results. In this work, we are investigating this theory on a larger data set, trying to find appropriate distance ranges for our model, facilitating future learning of qualitative representations of such proxemics thresholds.

Since the beginning of HRSI, the concept of proxemics is widely used and investigated in the field of social robotics. On the one hand, for human-aware navigation, many works adopt the zones defined by Hall [8] to achieve socially acceptable avoidance manoeuvres as can be seen from, e.g., [16] and most of the works on social cost functions listed below. On the other hand, there is the investigation of the optimal approach distance for a robot like the work by Torta *et al.* [19]. In their experiment, they investigate the optimal approach distance and angle for communication between a small humanoid robot [20] and a sitting person. They present an attractor based navigation framework that includes the definition for a *Region of Approach* which is optimal to communicate between the two agents. In the conducted experiment, Torta *et al.* show that an approach from the front is preferable over an approach from the side and found that the distance loosely correlates with the close phase of the social distance as defined by Hall [8]. Another example, focusing on the long-term habituation effects of approach distances is the work by Walters *et al.* [21]. They use a standing participant and a mobile service robot in an otherwise similar experimental setting as Torta *et al.* and inspect the long-term effect on the most comfortable approach distance. In our work we follow the same approach of investigating the optimal distances for our type of interaction but do not use self-assessment like Torta *et al.* or Walters *et al.* but a moving human and the recorded trajectories from our experiments. However, the presented work is meant to introduce a computational model and not to make assumptions about the quality of the actual interactions. We therefore assume that participants will keep the appropriate distance to the interaction partner and do not investigate the experienced comfort explicitly.

2.2. Path Planning

Path planning for mobile robots aims at finding a safe and short path which, in the majority of cases, is done by some form of A * algorithm. HRSI, on the other hand, does not aim to find the shortest or most energy efficient path but tries to adhere to numerous social norms and conventions, like the previously introduced proxemics [8], and thereby arguably makes navigation in human-populated environments safer and more efficient. There are several forms of human-aware path planning, using different forms of interaction representations of which examples will be given in the following.

One of the most common forms of representing humans in the environment is by using specific cost functions or potential fields, mainly circular or elliptical Gaussians [4,22–26]. These are used in the majority of human-aware path planners, employing a standard cost minimisation policy or more advanced planning algorithms like rapidly-exploring random trees (RRT) [25,26]. These approaches all rely on constraints or observed interactions and represent previous encounters via definitions to create or tune cost functions and potential fields rather than learning actual trajectories. Hence, they are a form of representing knowledge about human obstacles rather than representing previous interactions. This gives them the power to use generic path planners to create human-aware trajectories but deprives them of the wealth of information about the actual unfolding of such interactions stored in QTC state chains.

Other, less frequently used forms of representing HRSI for path planning include Social Force Models, Trajectory Learning, Heat Maps, or Motion Primitives. Social Forces have been used to describe inter-group relations and the drive of a human or a group of humans towards a goal, passing several subgoals or avoiding obstacles [27,28] which can also be transferred to robots to create more human-like behaviour. Social forces are therefore a way of abstracting from actual trajectories, using mathematical formulations but is still based in metric space and does not represent previous encounters unlike the presented calculus.

Trajectory learning is one of the more closely related approaches to our QTC based approach in HRSI. Feil *et al.* [29] used Gaussian Mixture Models created from observed trajectories to abstract from the concrete metric representation whereas Garrido *et al.* [30] used Hidden Markov Models and trajectory key points. Both of these approaches use different forms of abstraction to create a general model for HRSI but are still relying on a metric representation and are therefore very environment dependent. Heat Maps are another form of abstraction that still focuses on metric space. Arvunin *et al.* [31] used recorded trajectories of humans approaching an experimenter to create a so-called "Value Map" which can be used to represent the most commonly used paths for a specific configuration. A different form of abstraction is representing metric space via grid cells or a lattice as done by Kushleyev *et al.* [32], which allowed them to represent interaction in a dynamic system by a so-called time-bound lattice, using motion primitives. This interesting approach however, has only been employed for multi-robot environments and never in HRSI.

All the representations previously or currently used in HRSI path planning are based on metric space and Cartesian coordinates whereas our probabilistic QTC model abstracts from the actual coordinate system, environment, and metric space by representing qualitative states that both agents passed through in order to achieve the observed interaction. This naturally allows to incorporate the humans actions into the robot's path planning and decision making which, according to Ducourant *et al.* [7], is a very important factor in Human-Human Interaction.

2.3. Prediction

One major advantage of having a representation of interactions between a robot and a human is to be able to predict possible interactions based on experience and not only rely on a reactive path planner. There are two major principles of prediction used in HRSI: Prediction based on *reasoning* and on *learning* [9].

Prediction based on geometric reasoning follows constraints in the usual movement of humans given a certain environment and obstacles. Tadokoro *et al.* [33] use grid cells with an assigned probability–according to previous observations–of possible state transitions, meaning the likelihood of a human moving from one grid cell to the other. Ohki *et al.* [34] presented a similar approach also based on grid cells and their transition probability derived from the personal space of the human. Both of these approaches represent HRSI via state transitions but only focus on the possible future paths of the human and not the actual interaction between the robot and the human, unlike our approach. There are many other approaches that make assumptions about the future movement of humans given obstacles or certain environments which do not focus on HRSI but on preventing it by planning routes avoiding humans and are therefore not related to the presented QTC approach and will not be mentioned here.

Prediction based on learning means the collection of data and the creation of new samples from the built models which is highly related to the proposed probabilistic QTC representation but is currently almost exclusively based on map coordinates instead of abstract, qualitative states. Some of the more closely related works are on Motion Patterns, Feature Based Markov Decision Processes, and Short

Term Trajectory Libraries. Bennewitz *et al.* [35,36] use motion patterns as inputs for Hidden Markov Models to not only predict the immediate future state of a human during interaction but also possible trajectories the human takes through a previously observed office environment. Ziebart *et al.* [37] learn cost functions of the environment that explain previously observed behaviour and employ it in a Markov Decision process which enables them to plan paths that balance time-to-goal and pedestrian disruption in known and unknown environments. This transferral of knowledge is, due to its qualitative and abstract nature, also one of the main qualities of our QTC model. Chung *et al.* [38] observed pedestrians and created a library of short-term trajectories which they clustered to create pedestrian movement policies to predict how humans will move to avoid obstacles or each other.

All these approaches have in common that they not only map coordinates or trajectories to represent the interaction but also only represent the human side of it. Hence, all these models do not allow to predict how the robot's behaviour could influence humans behaviour during the interaction which is a crucial factor in HRSI. In contrast to all other approaches listed in this section, the QTC-based approach presented in the following, allows to abstract from metric space completely and absolutely by employing a qualitative representation. Moreover, this model, by providing information about the movement of the two agents in relation to one another, allows to make assumptions about how their spatial behaviour might influence each other during the interaction, based on previous observations.

The remainder of the paper is structured as follows. The variants of QTC used for the description of HRSI together with the multi variant QTC approach are described in Section 3. Our probabilistic sequential model utilising the described calculus is presented in Section 4. Section 5 shows the two experiments and methods used to evaluate this model, leading to the results shown in Section 6. Finally we are discussing the results in Section 7 and conclude in Section 8, showing some future work possibilities in Section 9.

3. The Qualitative Trajectory Calculus

In this section we will give an overview of the Qualitative Spatial Relation (QSR) we will use for our computational model. According to Kruse *et al.* [9], using QSRs for the representation of HRSI is a novel concept which is why we will go into detail about the two used versions of the calculus in question and also how we propose to combine them. This combination is employed to model distance thresholds implicitly using our probabilistic representation presented in Section 4.

The Qualitative Trajectory Calculus (QTC) belongs to the broad research area of qualitative spatial representation and reasoning [18], from which it inherits some of its properties and tools. The calculus was developed by Van de Weghe in 2004 to represent and reason about moving objects in a qualitative framework [15]. One of the main intentions was to enable qualitative queries in geographic information systems, but QTC has since been used in a much broader area of applications. Compared to the widely used Region Connection Calculus [39], QTC allows to reason about the movement of disconnected objects (DC), instead of unifying all of them under the same category, which is essential for HRSI. There are several versions of QTC, depending on the number of factors considered (e.g., relative distance, speed, direction, *etc.*) and on the dimensions, or constraints, of the space where the points move. The two most important variants for our work are QTC_B which represents movement in 1D and

QTC_C representing movement in 2D. QTC_B and QTC_C have originally been introduced in the definition of the calculus by Van de Weghe [15] and will be described to explain their functionality and show their appropriateness for our computational model. Their combination (QTC_{BC}) is an addition proposed by the authors to enable the distance modelling and to highlight the interaction of the two agents in close vicinity to one another. All three versions will be described in detail in the following (an implementation as a python library and ROS node can be found at [40]).

3.1. QTC Basic and QTC Double-Cross

The simplest version, called QTC Basic (QTC_B), represents the 1D relative motion of two points k and l with respect to the reference line connecting them (see Figure 1a). It uses a 3-tuple of qualitative relations (q_1 q_2 q_3), where each element can assume any of the values $\{-, 0, +\}$ as follows:

(q_1) movement of k with respect to l

> $-$: k is moving towards l
>
> 0 : k is stable with respect to l
>
> $+$: k is moving away from l

(q_2) movement of l with respect to k: as above, but swapping k and l

(q_3) relative speed of k with respect to l

> $-$: k is slower than l
>
> 0 : k has the speed of l
>
> $+$: k is faster than l

To create a more general representation we will use the simplified version QTC_{B11} which consists of the 2-tuple (q_1 q_2). Hence, this simplified version is ignorant of the relative speed of the two agents and restricts the representation to model moving *apart* or *towards* each other or being *stable* with respect to the last position. Therefore, the state set $S_B = \{(q_1, q_2) : q_1, q_2 \in \{-, 0, +\}\}$ for QTC_{B11} has $|S_B| = 3^2$ possible states and $|\tau_B| = |\{s \rightsquigarrow s' : s, s' \in S_B \wedge s \neq s'\}| = 32$ legal transitions as defined in the Conceptual Neighbourhood Diagram (CND). We are adopting the notation $s_1 \rightsquigarrow s_2$ for valid transitions according to the CND from [15], shown in Figure 2. By restricting the number of possible transitions–assuming continuous observations of both agents–a CND reduces the search space for subsequent states, and therefore the complexity of temporal QTC sequences.

(a)

(b)

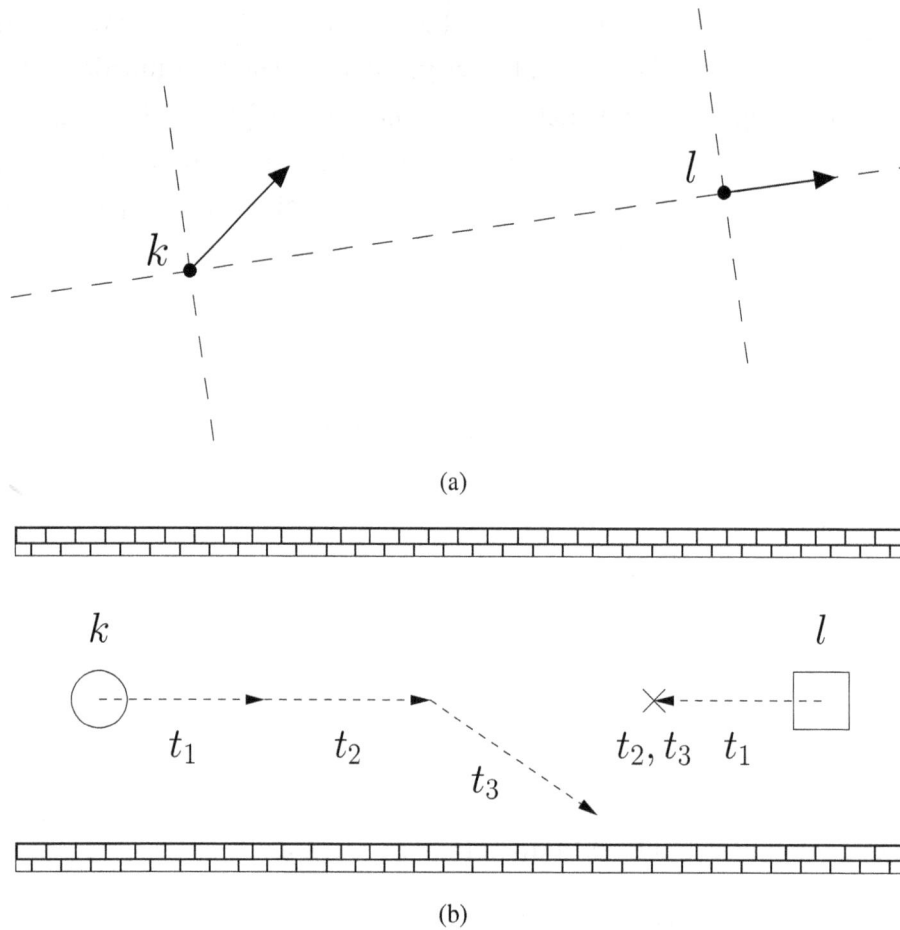

Figure 1. Example of moving points k and l. (**a**) The QTC_C double cross. The respective QTC_B and QTC_C relations for k and l are $(-+)$ and $(-+-0)$; (**b**) Example of a typical pass-by situation in a corridor. The respective QTC_C state chain is $(-- \ \ 0 \ 0)_{t_1} \rightsquigarrow (-0\,0\,0)_{t_2} \rightsquigarrow (-0+0)_{t_3}$.

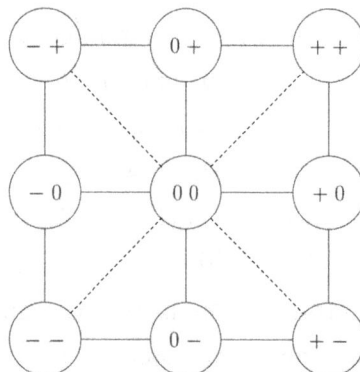

Figure 2. Conditional Neighbourhood Diagramm (CND) of QTC_{B11}. Given continuous observation, it is impossible to transition from moving towards the other agent to moving away from it without passing through the 0-state. Hence, whereas $+$ and $-$ always described intervals in time, 0 states can be of infisitinimal length. Also, note that due to the original formulation [15], there are no direct transitions in the CND between some of the states that, at a first glance, appear to be adjacent (e.g., (-0) and $(0-)$).

The other version of the calculus used in our models, called QTC Double-Cross (QTC$_C$) for 2D movement, extends the previous one to include also the side the two points move to, *i.e.*, left, right, or straight, and the minimum absolute angle of k, again with respect to the reference line connecting them (see Figure 1a). Figure 1b shows an example human-robot interaction in a corridor, encoded in QTC$_C$. In addition to the 3-tuple $(q_1\ q_2\ q_3)$ of QTC$_B$, the relations $(q_4\ q_5\ q_6)$ are considered, where each element can assume any of the values $\{-, 0, +\}$ as follows:

(q_4) movement of k with respect to $\overrightarrow{k\,l}$

> $-\ :\ k$ is moving to the left side of $\overrightarrow{k\,l}$
>
> $0\ :\ k$ is moving along $\overrightarrow{k\,l}$
>
> $+\ :\ k$ is moving to the right side of $\overrightarrow{k\,l}$

(q_5) movement of l with respect to $\overrightarrow{l\,k}$: as above, but swapping k and l

(q_6) minimum absolute angle of k, α_k with respect to $\overrightarrow{k\,l}$

> $-\ :\ \alpha_k < \alpha_l$
>
> $0\ :\ \alpha_k = \alpha_l$
>
> $+\ :\ \alpha_k > \alpha_l$

Similar to QTC$_B$ we also use the simplified version of QTC$_C$, QTC$_{C21}$. For simplicity we will from here on refer to the simplified versions of QTC, *i.e.*, QTC$_{B11}$ and QTC$_{C21}$ [15], as QTC$_B$ and QTC$_C$ respectively. This simplified version inherits from QTC$_B$ the ability to model if the agents are moving *apart* or *towards* each other or are *stable* with respect to the last position and in addition is also able to model to which side of the connecting line the agents are moving. The resulting 4-tuple $(q_1\ q_2\ q_4\ q_5)$ representing the state set $S_C = \{(q_1, q_2, q_4, q_5) : q_1, q_2, q_4, q_5 \in \{-, 0, +\}\}$, has $|S_C| = 3^4$ states, and $|\tau_C| = |\{s \rightsquigarrow s' : s, s' \in S_C \wedge s \neq s'\}| = 1088$ legal transitions as defined in the corresponding CND [41], see Figure 3.

These are the original definitions of the two used QTC variants which can be used in our computational model to identify HRSI encounters as shown in previous work [13]. To model distance however, we need both, QTC$_B$ and QTC$_C$, in one unified model. As shown in [12], QTC$_B$ and QTC$_C$ can be combined using hand crafted and simplified state chains and transitions to represent and reason about HRSIs. In the following section, however, we formalise and automatise this process, ultimately enabling us to use real world data to learn the transitions between the two variants of QTC instead of predefining them *manually*.

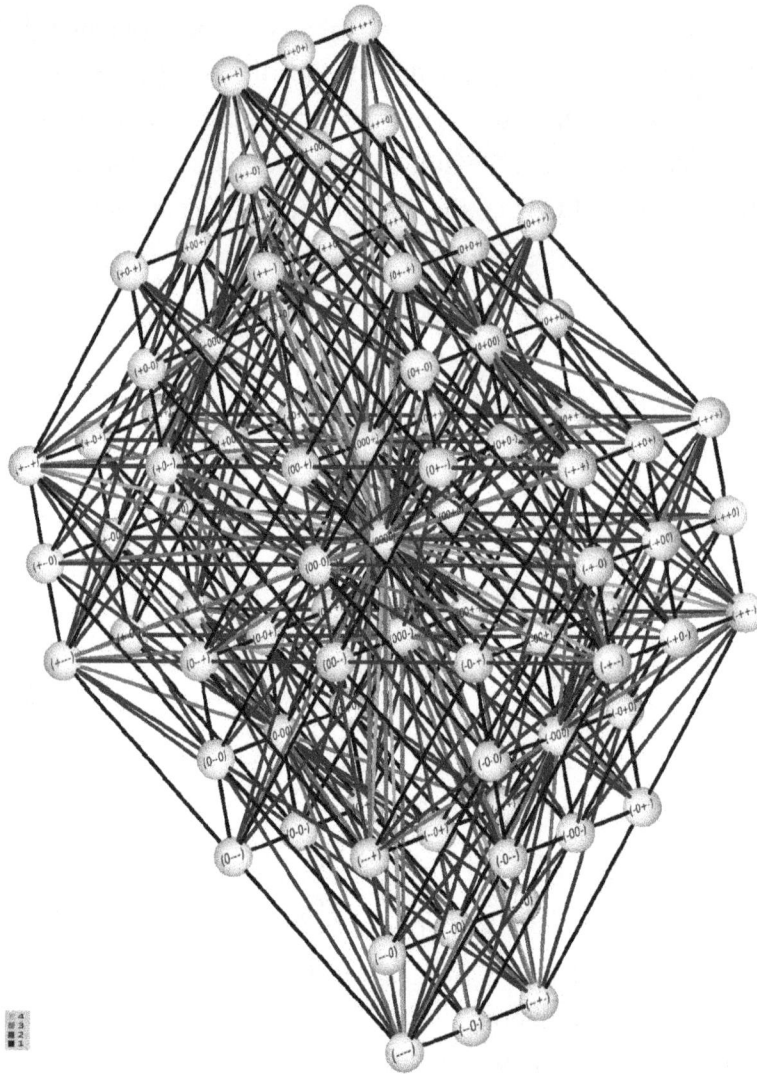

Figure 3. CND of the simplified QTC_C (image source [41]). Note that, similar to the CND for QTC_B, due to the original formulation of CNDs, there are no direct transitions between some of the states that at first glance look adjacent, e.g., $(-0+-)$ and $(-+0-)$.

3.2. Combining QTC_B and QTC_C

As mentioned, one of the aims of the proposed computational model is the combination of the 1D representation QTC_B, used when the two agents are far apart, and the 2D representation QTC_C, used when the agents are in close vicinity to one another. To achieve this combination via the implicit modelling of the distance threshold used and the resulting simplification of QTC state chains for HRSI when encapsulated in our probabilistic framework, we propose the combination of QTC_B and QTC_C referring to it as QTC_{BC}. The combined variants are QTC_{B11} and QTC_{C21} which results in QTC_{B11C21}, from here on referred to as QTC_{BC} for simplicity. This proposed model is in principal able to switch between the fine and coarse version of QTC at will, but for the presented work we will use distance thresholds to trigger the switching. These thresholds could be social distances like the previously mentioned area of far phase personal space and close phase social space as defined by Hall [8] but could also be any other distance value. Since the actual quantitative distance threshold used is not explicitly included in the QTC_{BC} tuple, it will be modelled implicitly via the transition between the two enclosed variants.

The set of possible states for QTC_{BC} is a simple unification of the fused QTC variants. In the presented case the integrated QTC_{BC} states are defined as:

$$S_I = S_B \cup S_C \tag{1}$$

with $|S_I| = |S_B| + |S_C| = 90$ states.

The transitions of QTC_{BC} include the unification of the transitions of QTC_B and QTC_C–as specified in the corresponding CNDs (see Figures 2 and 3)–but also the transitions from QTC_B to QTC_C: $\tau_{BC} = \{s_b \rightsquigarrow s_c : s_b \in S_B, s_c \in S_C\}$ and from QTC_C to QTC_B: $\tau_{CB} = \{s_c \rightsquigarrow s_b : s_b \in S_B, s_c \in S_C\}$, respectively. This leads to the definition of QTC_{BC} transitions as:

$$\tau_I = \tau_B \cup \tau_C \cup (\tau_{BC} \cup \tau_{CB}) \tag{2}$$

To preserve the characteristics and benefits of the underlying calculus τ_{BC} and τ_{CB} are simply regarded as an increase or decrease in granularity, *i.e.*, switching from 1D to 2D or vice-versa. As a result there are two different types of transitions:

1. Pseudo self-transitions where the values of $(q_1 \, q_2)$ do not change, plus all possible combinations for the 2-tuple $(q_4 \, q_5)$: $|S_B| \cdot 3^2 = 81$, e.g., $(++) \rightsquigarrow (++--)$ or $(++--) \rightsquigarrow (++)$.
2. Legal QTC_B transitions, plus all possible combinations for the 2-tuple $(q_4 \, q_5)$: $|\tau_B| \cdot 3^2 = 288$, e.g., $(+0) \rightsquigarrow (++--)$ or $(+0--) \rightsquigarrow (++)$.

Resulting into:

$$|\tau_{BC}| + |\tau_{CB}| = 2 \cdot (81 + 288) = 738$$

transitions between the two QTC variants. This leads to a total number of QTC_{BC} transitions of:

$$\begin{aligned} \tau_I &= |\tau_B| + |\tau_C| + (|\tau_{BC}| + |\tau_{CB}|) \\ &= 32 + 1088 + 738 \\ &= 1858 \end{aligned}$$

These transitions depend on the previous and current Euclidean distance of the two points $d(k, l)$ and the threshold d_s representing an arbitrary distance threshold:

$$\tau_I = \begin{cases} \tau_B & \text{if } d(k,l)_{t-1} > d_s \wedge d(k,l)_t > d_s, \\ \tau_{BC} & \text{else if } d(k,l)_{t-1} > d_s \wedge d(k,l)_t \leq d_s, \\ \tau_{CB} & \text{else if } d(k,l)_{t-1} \leq d_s \wedge d(k,l)_t > d_s, \\ \tau_C & \text{otherwise} \end{cases} \tag{3}$$

These transitions, distances, and threshold d_s play a vital role in our probabilistic representation of QTC_{BC} which will be described in the following section.

4. Probabilistic Model of State Chains

After introducing our model of QTC$_{BC}$ in Section 3, we will describe a probabilistic model of QTC$_{BC}$ state chains in the following. This probabilistic representation is able to learn QTC state chains and the transition probabilities between the states from observed trajectories of human and robot, using the distance threshold d_s to switch between the two QTC variants during training. This model is later on used as a classifier to compare different encounters and to make assumptions about the quality of the representation. This representation is able to compensate for illegal transitions and shall also be used in future work as a knowledge base of previous encounters to classify and predict new interactions.

In previous work, we proposed a probabilistic model of state chains, using a Markov Model and QTC$_C$ to analyse HRSI [11]. This first approach has been taken a step further and evolved into a Hidden Markov Model (HMM) representation of QTC$_C$ [13]. This enables us to represent actual sensor data by allowing for uncertainty in the recognition process. With this approach, we are able to reliably classify different HRSI encounters, e.g., head-on (see Figure 4) and overtake–where the human is overtaking the robot while both are trying to reach the same goal–scenarios, and show in Section 6 that the QTC-based representations of these scenarios are significantly different from each other.

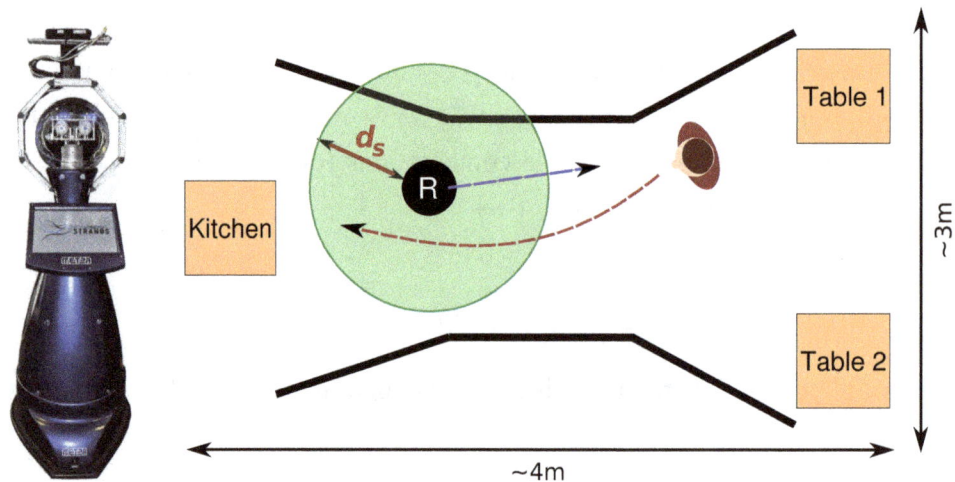

Figure 4. *Left:* Robot. Hight: 1.72 m, diameter: $\sim$ 61 cm. *Right:* Head-on encounter. Robot ("R") tries to reach a table while the human (reddish figure) is trying to reach the kitchen. Experimental set-up: kitchen on the left and two tables on the right. Black lines represent the corridor. Circle around robot represents a possible distance threshold d_s.

To be able to represent distance for future extensions of the HMM as a generative model, to highlight events in close vicinity to the human, and to create a more concise and tractable model, we propose the probabilistic representation of QTC$_{BC}$ state chains, using a similar approach as in [13]. We are now modelling the proposed QTC$_{BC}$ together with QTC$_B$ and QTC$_C$ which allows to dynamically switch between the two combined variants or to use the two pure forms of the calculus. This results in extended transition and emission probability matrices for τ_I (see Figure 5, showing the transition probability matrix) which incorporate not only QTC$_B$ and QTC$_C$, but also the transitional states defined by QTC$_{BC}$.

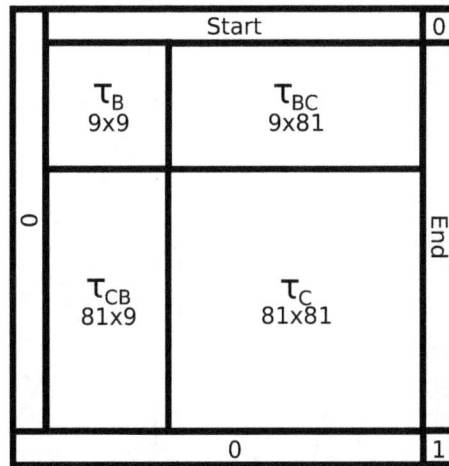

Figure 5. The HMM transition matrix τ_I for QTC$_{BC}$ as described in Equations (2) and (3).

Similar to the HMM based representation described in [13], we have initially modelled the "correct" emissions, e.g., $(+-)$ actually emits $(+-)$, to occur with 95% probability and to allow the model to account for detection errors with 5%. Our HMM contains $|\tau_I| + 2 \cdot |S_I| = 1858 + 2 \cdot 90 = 2038$ legal transitions stemming from QTC$_{BC}$ and the transitions from and to the start and end state, respectively (see Figure 5).

To represent different HRSI behaviours, the HMM needs to be trained from the actual observed data (see Figure 6, showing an example of a trained state chain using pure QTC$_C$). For each different behaviour to be represented, a separate HMM is trained, using Baum-Welch training [42] (Expectation Maximisation) to obtain the appropriate transition and emission probabilities for the respective behaviour. In the initial pre-training model, the transitions that are *valid*, according to the CNDs for QTC$_B$ and QTC$_C$ and our QTC$_{BC}$ definition for transitions between the two, are modelled as equally probable (uniform distribution). We allow for pseudo-transitions with a probability of $P_{pt} = 1e^{-10}$ to overcome the problem of a lack of sufficient amounts of training data and unobserved transitions therein. To create the training set we have to transform the recorded data to QTC$_C$ state chains that include the Euclidean distance between k and l and define a threshold d_s at which we want to transition from QTC$_B$ to QTC$_C$ and vice-versa (setting $d_s = 0$ results in a pure QTC$_B$ model and setting $d_s = \inf$ results in pure QTC$_C$ model–identical to the definition in [13] for full backwards compatibility). Of course, the actual values for d_s can be anything from being manually defined, taken from observation, or being a probabilistic representation of a range of distances at which to transition from one QTC variant to the other. For the sake of our evaluation we are showing a range of possible values for d_s in Section 6 to find suitable candidates.

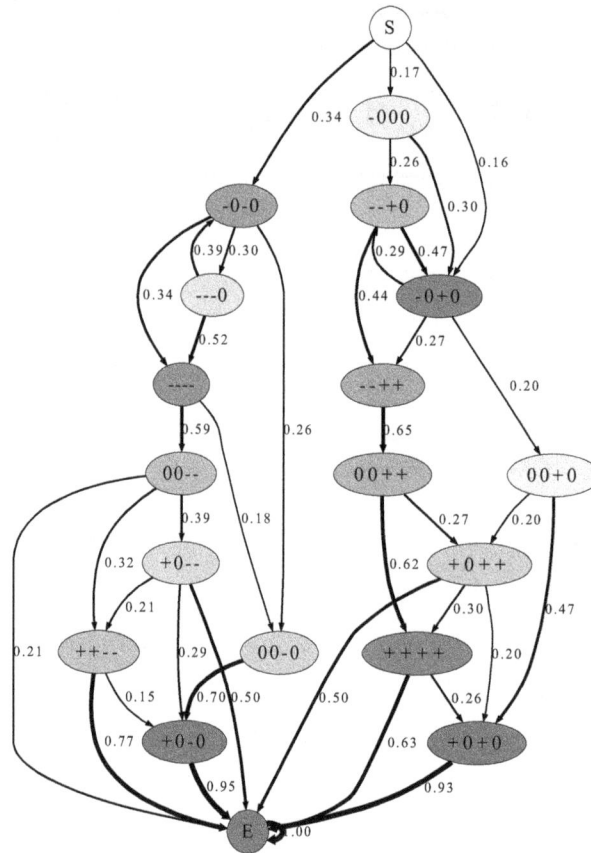

Figure 6. QTC$_C$ states for pass-by situations created by the HMM representation. Edge width represents the transition probability. The colour of the nodes represents the a-priori probability of that specific state (from white = 0.0, e.g., "S", to dark grey = 1.0, e.g., "E"). All transition probabilities below 0.15 have been pruned from the graph, only highlighting the most probable paths within our model. Due to the pruning, the transition probabilities in the graph do not sum up to 1.0.

To create a state chain similar to the exemplary one shown in Figure 7, the values for the side movement $(q_4\,q_5)$ of the QTC$_C$ representation are simply omitted if $d(k,l) > d_s$ and the remaining $(q_1\,q_2)$ 2-tuple for the 1D movement will be represented by the QTC$_B$ part of the transition matrix. If the distance crosses the threshold, it will be represented by one of the τ_{BC} or τ_{CB} transitions. The full 2D representation of QTC$_C$ is used in the remainder of the cases. Afterwards, all distance values are removed from the representation because the QTC state chain now implicitly models d_s, and similar adjacent states are collapsed to create a valid QTC representation (see Figure 7 for a conceptual state chain). This enables us to model distance via the transition between the QTC variants, while still using the pure forms of the included calculi in the remainder of the cases, preserving the functionality presented in [13], which will be shown in Section 6.

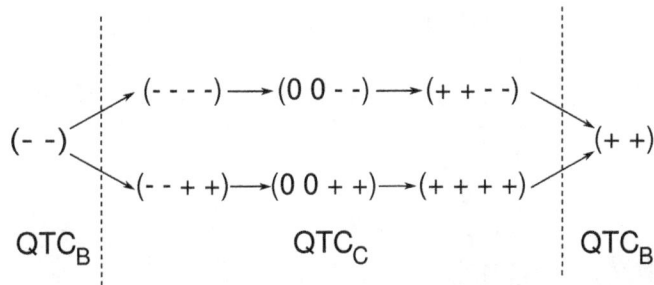

Figure 7. Conceptual temporal sequence of QTC_{BC} for a head-on encounter. From left to right: approach, pass-by on the left or right side, moving away. Dashed lines represent instants where the distance threshold d_s is crossed.

5. Experiments

To evaluate the soundness and representational capabilities of our probabilistic model of HRSI using QTC, particularly QTC_{BC}, state chains, we train our HMM representation using real-world data from two experiments. These HMMs are then employed as classifiers to generate a comparative measurement enabling us to make assumptions about the quality of the model and the distance thresholds d_s. The two experiments both investigate the movement of two agents in confined shared spaces. The first experiment, later referred to as "Lincoln experiment", was originally described in our previous work on QTC [13] and features a mobile service robot and a human naïve to the goal of the experiment. The tasks were designed around a hypothetical restaurant scenario eliciting incidental and spontaneous interactions between human and robot.

The second experiment, later referred to as the "Bristol experiment", features two agents (both human) passing each other in a two meter wide corridor. One of the two (the experimenter) was dressed up as a "robot", masking her body shape, and her face and eyes were hidden behind goggles and a face mask (see Figure 8). This "fake robot" received automated instructions on movement direction and collision avoidance strategy via headphones. Similar to the "Lincoln experiment", the other person was a participant naïve to the goal of the experiments, but has been given explicit instructions to cross the corridor with as little veering as possible, but without colliding with the oncoming agent. This second experiment does not feature a real robot but, yields similar results using our model, as can be seen in Section 6. Both experiments feature two agents interacting with each other in a confined shared space and are well suited to demonstrate the representational capabilities of our approach, showing how the approach can be effectively generalised or extended to other forms of spatial interaction.

Figure 8. The "Bristol Experiment" set-up. Corridor from the participants perspective before the start of a trial. Middle: experimenter dressed as "robot". The visual marker was attached to the wall behind the "robot" above her head.

In the following sections we will describe the general aims and outlines of the experiments used. This is meant to paint the bigger picture of the underlying assumptions and behaviours of the robot/experimenter during the interactions and to explain some of the conditions we compared in our evaluation. Both experiments investigated different aspects of HRSI and spatial interaction in general, which created data well suited for our analysis of the presented probabilistic model utilising QTC_{BC} and to investigate appropriate distance thresholds d_s.

5.1. "Lincoln Experiment"

This section presents a brief overview of the "Lincoln experiment" set-up and tasks. Note, the original aim of the experiment, besides the investigation of HRSI using an autonomous robot in general, was finding hesitation signals in HRSI [43], hence the choice of conditions.

5.1.1. Experiment Design

In this experiment the participants were put into a hypothetical restaurant scenario together with a human-size robot (see Figure 4). The experiment was situated in a large motion capture lab surrounded by 12 motion capture cameras (see Figure 9), tracking the x, y, z coordinates of human and robot with a rate of 50 Hz and an approximate error of 1.5 mm~2.5 mm. The physical set-up itself was comprised of two large boxes (resembling tables) and a bar stool (resembling a kitchen counter). The tables and the kitchen counter were on different sides of the room and connected via a ~2.7 m long and ~1.6 m wide artificial corridor to elicit close encounters between the two agents while still being able to reliably track their positions (see Figure 4). The length is just the length of the actual corridor, whereas the complete set-up was longer due to the added tables and kitchen counter plus some space for the robot and human to turn. The width is taken from the narrowest point. At the ends, the corridor widens to ~ 2.2 m to give more room for the robot and human to navigate as can be seen in Figure 4. The evaluation however will only regard interactions in this specified corridor. For this experiment we had 14 participants (10 male, 4 female) who interacted with the robot for 6 minutes each. All of the participants were employees or students at the university and 9 of them have a computer science background; out of these 9 participants only 2 had worked with robots before. The robot and human were fitted with motion capture markers to track their x, y coordinates for the QTC representation–Figure 10 shows an example of recorded trajectories (the raw data set containing the recorded motion capture sequences is publicly available on our git repository [44]).

The robot was programmed to move autonomously back and forth between the two sides of the artificial corridor (kitchen and tables), using a state-of-the-art planner [45,46]. Two different behaviours were implemented, *i.e.*, *adaptive* and *non-adaptive* velocity control, which were switched at random ($p = 0.5$) upon the robot's arrival at the kitchen. The adaptive velocity control gradually slowed down the robot, when entering the close phase of the social space [8], until it came to a complete stand still before entering the personal space [8] of the participant. The non-adaptive velocity control ignored the human even as an obstacle (apart from an emergency stop when the two interactants were too close, approx. <0.4 m, to prevent injuries), trying to follow the shortest path to the goal, only regarding static obstacles. This might have yielded invalid paths due to the human blocking it, but led to the desired robot behaviour of not respecting the humans personal space. We chose to use these two distinct behaviours because they mainly differ in the speed of the robot and the distance it keeps to the human. Hence, they produce very similar, almost straight trajectories which allowed us to investigate the effect of distance and speed on the interaction while the participant was still able to reliably infer the robot's goal. As mentioned above, this was necessary to find hesitation signals [43].

Figure 9. The "Lincoln experiment" set-up showing the robot, the motion capture cameras, the artificial corridor, and the "tables" and "kitchen counter". The shown set-up elicits close encounters between human and robot in a confined shared space to investigate their interaction.

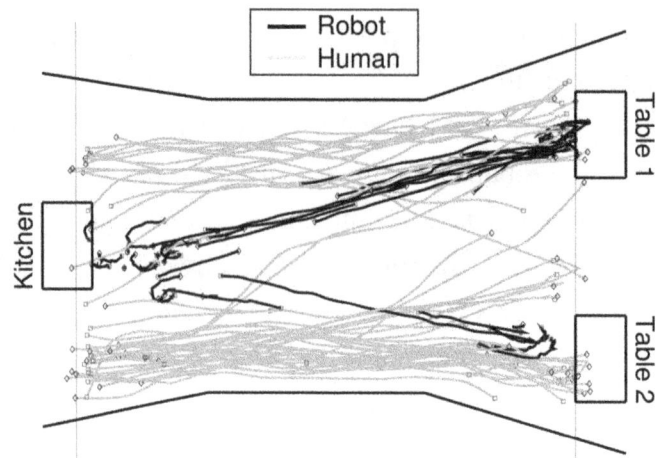

Figure 10. The recorded trajectories of one of the participants (grey = human, black = robot). The rough position of the corridor walls and the furniture is also depicted. The pink lines on either side show the cut-off lines for the evaluation. The robots trajectories were not bound to the cut-off lines but to the humans trajectories timestamps. The humans trajectories themselves might not end at the cut-off line but before due to those regions being on the outside limits of the tracking region, causing the loss of markers by the tracking system.

Before the actual interaction, the human participant was told to play the role of a waiter together with a robotic co-worker. This scenario allowed to create a natural form of pass-by interaction (see Figures 4 and 1b) between human and robot by sending the participants from the kitchen counter to the tables and back to deliver drinks, while at the same time the robot was behaving in the described way. This task only occasionally resulted in encounters between human and robot but due to the incidental nature of these encounters and the fact that the participants were trying to reach their goal as efficiently as possible, we hoped to achieve a more natural and instantaneous participant reaction.

5.2. "Bristol Experiment"

In the following, we will give a quick overview of the "Bristol experiment" set-up and tasks. Besides investigating general HRSI concepts, the main aim of the experiment was to investigate the impact and dynamics of different visual signal types to inform an on-coming agent of the direction of intended avoidance manoeuvres in an artificial agent in HRSI, hence the comparatively complex set-up of conditions. For the purpose of the QTC analysis presented, however, we will just look at a specific set of conditions out of the ones mentioned in the experiment description.

5.2.1. Experiment Design

In this experiment, 20 participants (age range 19–45 years with a mean age of 24.35) were asked to pass an on-coming "robotic" agent (as mentioned above, a human dressed as a robot, from now on referred to as "robot") in a wide corridor. The corridor was placed in the Bristol Vision Institute (BVI) vision and movement laboratory, equipped with 12 Qualisys 3D-motion capture cameras. The set-up allows to track movement of motion capture markers attached to the participants and the robot in

x, y, z-coordinates over an area of 12 m (long) $\times$ 2 m (wide) $\times$ 2 m (high) (see Figure 8) with a frequency of 100 Hz and an approximate error of 1 mm.

Participants were asked to cross the laboratory toward a target attached to the centre of the back wall (and visible at the beginning of each trial at the wall above the head of the "robot") as directly as possible, without colliding with the on-coming "robot". At the same time, the "robot" would cross the laboratory in the opposite direction, thus directly head-on to the participant. In 2/3 of the conditions, the "robot" would initiate an automated "avoidance behaviour" to the left or right of the participant that could be either accompanied by a visual signal indicating the direction of the avoidance manoeuvre or be unaccompanied by visual signals (see Figure 11 for the type of signals). Note that if neither robot nor participant were to start an avoidance manoeuvre, they would collide with each other approximately midway through the laboratory.

The robot, dressed in a black long-sleeved T-shirt and black leggings, was wearing a "robot suit" comprising of two black cardboard boards (71 cm high $\times$ 46 cm wide) tied together over the agent's shoulders on either side with belts (see Figure 8). The suit was intended to mask body signals (e.g., shoulder movement) usually sent by humans during walking. To also obscure the "robots" facial features and eye gaze, the robot further wore a blank white mask with interiorly attached sunglasses.

A Nexus 10 Tablet (26 cm $\times$ 18 cm) was positioned on the cardboard suit at chest height to display a "go" signal at the beginning of each trial to inform the participant that they should start walking. The go signal was followed 1.5 s later by the onset of visual signals (cartoon eyes, indicators, or a blank screen as "No signal"). With exception of the "no signal", these visual signals stayed unchanged in a third of the trials, and in the other 2 thirds of trials, they would change 0.5 s later to signal the direction in which the robot would try to circumvent the participant (the cartoon eyes would change from straight ahead to left or right, the indicators would start flashing left or right with a flash frequency of 2 Hz). Note that no deception was used; *i.e.*, if the robot indicated a direction to the left or right, it would always move in this direction. However, if the robot did not visually indicate a direction, it would still move to the left or right in two thirds of trials. Only in the remaining trials, the robot would keep on walking straight, thus forcing the participant to avoid collision by actively circumventing the robot.

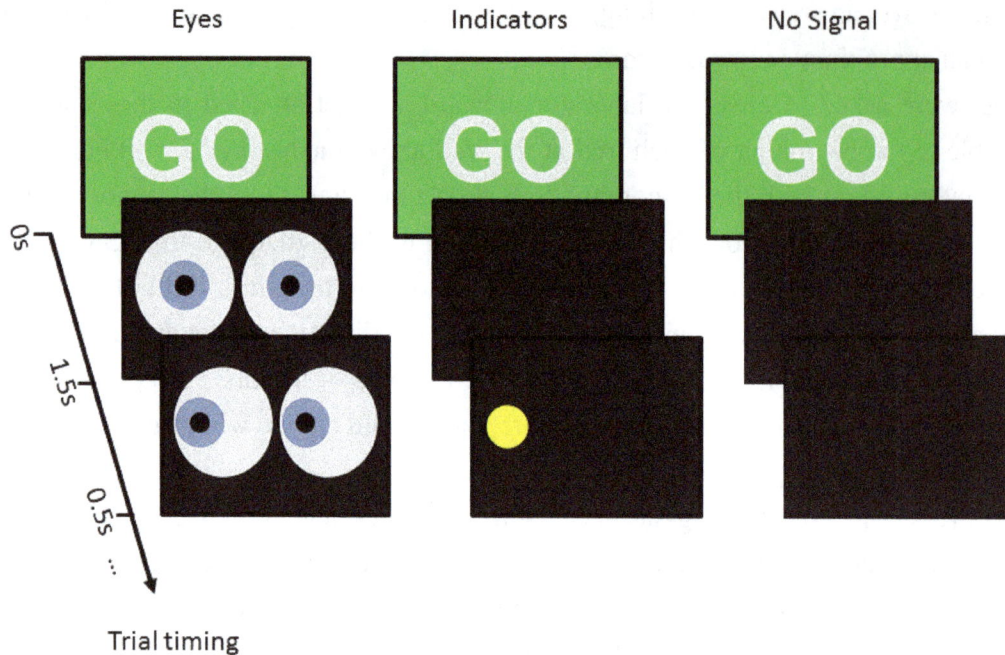

Figure 11. Examples of visual signals sent by the "robot" agent. Visual signal onset occurred 1.5 s after the "go" signal encouraging the participant to start walking. 500 ms after the signal onset, signals could either then change to indicate a clear direction in which the robot would avoid the participant or remain uninformative with respect to the direction of movement of the robot.

The actual/physical onset of the robot's avoidance manoeuvres could start 700 ms before the visual directional signal was given (early), at the time of the visual direction signal (middle), or 700 ms after the onset of the visual direction signal (late). These three conditions will later on be referred to as *early*, *middle*, or *late*, respectively.

In the following we will focus on the trajectories taken by both agents in this confined shared space and compare some of the conditions used.

5.3. Evaluation

The aim of the evaluation is to test the descriptive quality of the created probabilistic sequential model utilising QTC state chains in general, to evaluate possible distance thresholds or ranges of thresholds to be incorporated into the model, and to learn appropriate transitions between the QTC variants for our QTC_{BC} model. To this end, the models created from the recorded trajectories are employed as classifiers to generate comparative measurements, allowing to make statements about the representational quality of the model itself. These classifiers use a range of distance thresholds to find those values appropriate for the switch from QTC_B to QTC_C and vice-versa. The goal of this evaluation therefore is not to compare the quality or appropriateness of the different QTC variants but their combination with our probabilistic model to represent HRSI. Hence, classification is not only an important application for our model but we are also using it as a tool to create a comparative measurement for evaluation.

To quickly recapitulate, the different QTC variants we are using in the following are: (i) QTC_B-*1D*: represents approach, moving away, or being stable $(-+0)$ in relation to the last position; (ii) QTC_C-*2D*:

in addition to QTC_B, also includes to which side the agents are moving, left of, right of, or along $(- + 0)$ the connecting line; (iii) *QTC_{BC}-1D/2D:* the combination of both according to the distance of the two agents, QTC_B (1D) when far apart and QTC_C (2D) when close. The 0 states mentioned in the following are therefore instances in time when the agent was stable in its 1-dimensional and/or 2-dimensional movement.

The data of both experiments will be used equally for our evaluation. However, due to the different nature of the investigated effects and signals, and the resulting different set-ups used, there will be slight differences in the evaluation process and therefore it will be split in two parts according to the experiments. The used model on the other hand, will be the same for both experiments to show its generalisability. In the following we will present the used evaluation procedures for each study.

Lincoln Experiment   We defined two virtual cut-off lines on either side of the corridor (see Figure 10) to separate the trajectories into trials and because we only want to investigate close encounters between human and robot and therefore just used trajectories inside the corridor. Out of these trajectories, we manually selected 71 head-on and 87 overtaking encounters and employed two forms of noise reduction on the recorded data. The actual trajectories were smoothed by averaging over the x, y coordinates for 0.1 s, 0.2 s, and 0.3 s. The z coordinate is not represented in QTC. To determine 0 QTC states–one or both agents move along $\overrightarrow{k\,l}$ or along the two perpendicular lines (see Figure 1)–we used three different quantisation thresholds: 1 cm, 5 cm, and 10 cm, respectively. Only if the movement of one or both of the agents exceeded these thresholds it was interpreted as a $-$ or $+$ QTC state. This smoothing and thresholding is necessary when dealing with discrete sensor data which otherwise would most likely never produce 0 states due to sensor noise.

To find appropriate distance thresholds for QTC_{BC}, we evaluated distances on a scale from pure QTC_B (40 cm) to pure QTC_C (3 m), in 10 cm steps. The $d_s < 0.4$ m threshold represents pure QTC_B because the robot and human are represented by their centre points, therefore, it is impossible for them to get closer than 40 cm . On the other hand, the $d_s \geq 3$ m threshold represents pure QTC_C because the corridor was only ~ 2.7 m long.

We evaluated the head-on *vs.* overtake, passing on the left *vs.* right, and adaptive *vs.* non-adaptive velocity conditions.

Bristol Experiment   Following a similar approach as described above, we split the recorded data into separate trials, each containing one interaction between the "robot" and the participant. To reduce noise caused by minute movements before the beginning and after the end of a trial, we removed data points from before the start and after the end of the individual trial by defining cut off lines on either end of the corridor, only investigating interactions in between those boundaries. Visual inspection for missing data points and tracking errors yielded 154 erroneous datasets out of the 1439 trials in total and were excluded from the evaluation. Similar to the Lincoln data set, we applied three different smoothing levels 0.00 s, 0.02 s, and 0.03 s. We also used four different quantisation levels, 0.0 cm, 0.1 cm, 0.5 cm, and 1 cm, to generate QTC 0-states (due to the higher recording frequency of 100 Hz the smoothing and quantisation values are lower than for the Lincoln experiment). Unlike the "Lincoln experiment", one of the smoothing and quantisation combinations, *i.e.*, 0.0 s and 0.0 cm, represents unsmoothed and unquantised data. This was possible due to a higher recording frequency and a less noisy motion capture system.

We evaluated distances on a scale from pure QTC_B (40 cm) to 3 m, in 10 cm steps. To stay in line with our first experiment, we evaluated distances of up to 3 m but since the corridor had a length of 12 m, we also added a pure QTC_C representation ($d_s = \text{inf}$) for comparison.

In this experiment we did not investigate overtaking scenarios as those were not part of the experimental design. We evaluated passing on the left vs. right and indicator vs. no indicator separated according to their timing condition (i.e., early, middle, late), and early vs. late regardless of any other condition.

Statistical Evaluation To generate the mentioned comparative measurement to evaluate the meaningfulness of the representation, we used our previously described HMM based QTC_{BC} representation as a classifier comparing different conditions. With this measurement, we are later on able to make assumptions about the quality and representational capabilities of the model itself.

For the classification process, we employed k-fold cross validation with $k = 5$, resulting in five iterations with a test set size of 20% of the selected trajectories. This was repeated ten times for the "Lincoln experiment" and 4 times for the "Bristol experiment"–to compensate for possible classification artefacts due to the random nature of the test set generation–resulting in 50 and 20 iterations over the selected trajectories, respectively. The number of repetitions for the "Bristol experiment" is lower due to the higher number of data points and the resulting increase in computation time and decrease in feasibility. Subsequently, a normal distribution was fitted over the classification results to generate the mean and 95% confidence interval and make assumptions about the statistical significance. Being significantly different from the null hypothesis (H_0; $p = 0.5$) for the evaluations presented in the following section would therefore imply that our model is expressive enough to represent the encounter it was trained for. This validation procedure was repeated for all smoothing and quantisation combinations.

6. Results

To verify the effectiveness of our probabilistic representation of QTC_{BC} state chains given different distance thresholds, we used the described classifiers to generate a comparative measure by evaluating the classification rate for our two experiments. We evaluated head-on vs. overtake and adaptive vs. non-adaptive velocity control in the "Lincoln experiment", passing on the left vs. passing on the right in both, and early vs. late and indicator vs. no indicator in the "Bristol experiment". Figure 7 shows a conceptual example of a resulting QTC_{BC} representation of a head-on encounter which is the most dominant in both experiments.

6.1. Results "Lincoln Experiment"

Table 1a shows the minimum and maximum classification rates (μ) for the general head-on vs. overtaking case and the respective QTC_{BC} thresholds (d_s). For the majority of the different smoothing levels (7 out of 9), the best classification results were achieved using distance thresholds of $QTC_B \leq d_s \leq 0.6$ m. The best result $\mu = 0.98$ was achieved using a distance of $d_s = 2.2$ m and smoothing values of 0.3 s and a quantisation value of 1 cm. Even though the lowest and highest classification rates for the different smoothing and quantisation levels are significantly different from each other, they are all significantly different from H_0 as well. The overall worst results have been achieved using a smoothing value of 0.1 s and a quantisation level of 10 cm. Using this combination yields the highest number of 0-states compared to all the other combinations due to the fact that for a movement to be recognised it has to diverge from the previous position by 10 cm which is very unlikely to happen in 0.1 s.

The comparison of passing on the left *vs.* passing on the right, is shown in Table 1b. All of the results show bad classification rates if $d_s \leq 0.7$ m, and high classification results for values of $d_s \geq 0.9$ m. Figure 12a shows two typical results from the "Lincoln experiment" using the lowest and highest smoothing levels. The higher smoothing and quantisation value combination, and the resulting reduced noise, show a steeper incline in classification rates than the lowest value combination, which can be seen from the smaller yellow area in the right half of Figure 12a. Nevertheless, in all of the cases, a sudden increase in performance (jumping from $\mu \approx 0.5$ to $\mu > 0.8$) can be seen at 0.9 m $\leq d_s \leq 1.2$ m.

The third case, adaptive *vs.* non-adaptive robot behaviour in head-on encounters, is shown in Table 1c. This behaviour did not result in different trajectories during the interaction but only differed in the time it took the robot to traverse the corridor. Due to the definition of QTC it is not able to represent absolute time, which makes it hard to classify these two behaviours accordingly. The best results for each quantisation level were achieved at distances of $QTC_B \leq d_s \leq 0.7$ m, all lying on the diagonal of Table 1c. Since time is a crucial factor in this condition, it is very dependent on the right smoothing value combination. Figure 12b shows two exemplary results. The left hand side depicts the best classification result with classification rates of up to $\mu = 0.748$ for $d_s = 0.7$ m. The right hand side shows the results for a smoothing level that did not yield the best results for low but medium distance threshold of $d_s = 1.5$ m with a classification rate of $\mu = 0.643$.

Table 1. Classification results "Lincoln experiment", **bold:** mentioned in text. The mentioned confidence intervals represent the boundary cases and all the others can be considered lower. (**a**) Head-on *vs.* Overtake. Maximum 95% confidence intervals ($p < 0.05$) for min and max classification results: *min:* 0.0209, *max:* 0.0182; (**b**) Head-on: Left *vs.* Right. Maximum 95% confidence intervals ($p < 0.05$) for min and max classification results: *min:* 0.0221, *max:* 0.0182; (**c**) Head-on: Adaptive *vs.* Non-Adaptive. Maximum 95% confidence intervals ($p < 0.05$) for min and max classification results: *min:* 0.0202, *max:* 0.0251.

(a)

Smoothing		0.1 s		0.2 s		0.3 s	
	Res.	μ	d_s	μ	d_s	μ	d_s
1 cm	min	0.90	0.7	0.89	1.0	0.91	0.7
	max	0.97	QTC$_C$	0.96	**0.6**	**0.98**	**2.2**
5 cm	min	0.84	0.8	0.88	0.8	0.87	0.7
	max	0.92	**0.5**	0.97	**QTC$_B$**	0.94	**QTC$_B$**
10 cm	min	**0.70**	2.0	0.79	1.2	0.79	0.9
	max	**0.82**	**QTC$_B$**	0.87	**0.5**	0.89	**0.4**

(b)

Smoothing		0.1 s		0.2 s		0.3 s	
	Res.	μ	d_s	μ	d_s	μ	d_s
1 cm	min	**0.50**	**QTC$_B$**	0.58	QTC$_B$	0.52	QTC$_B$
	max	**0.97**	**1.9**	0.95	2.4	0.96	2.3
5 cm	min	0.41	QTC$_B$	0.41	QTC$_B$	0.49	QTC$_B$
	max	0.90	2.9	0.93	2.8	0.94	2.9
10 cm	min	0.50	QTC$_B$	0.43	QTC$_B$	**0.52**	**0.5**
	max	0.92	QTC$_C$	0.90	1.2	**0.95**	**QTC$_C$**

(c)

Smoothing		0.1 s		0.2 s		0.3 s	
	Res.	μ	d_s	μ	d_s	μ	d_s
1 cm	min	**0.46**	**1.4**	0.48	1.8	0.47	0.5
	max	**0.66**	**QTC$_B$**	0.60	0.8	0.64	1.5
5 cm	min	0.52	1.0	**0.55**	**1.4**	0.54	1.3
	max	0.69	1.5	**0.75**	**0.7**	0.72	0.5
10 cm	min	0.46	1.2	0.49	0.8	**0.59**	**1.6**
	max	0.60	1.8	0.64	1.0	**0.74**	**0.7**

(a)

(b)

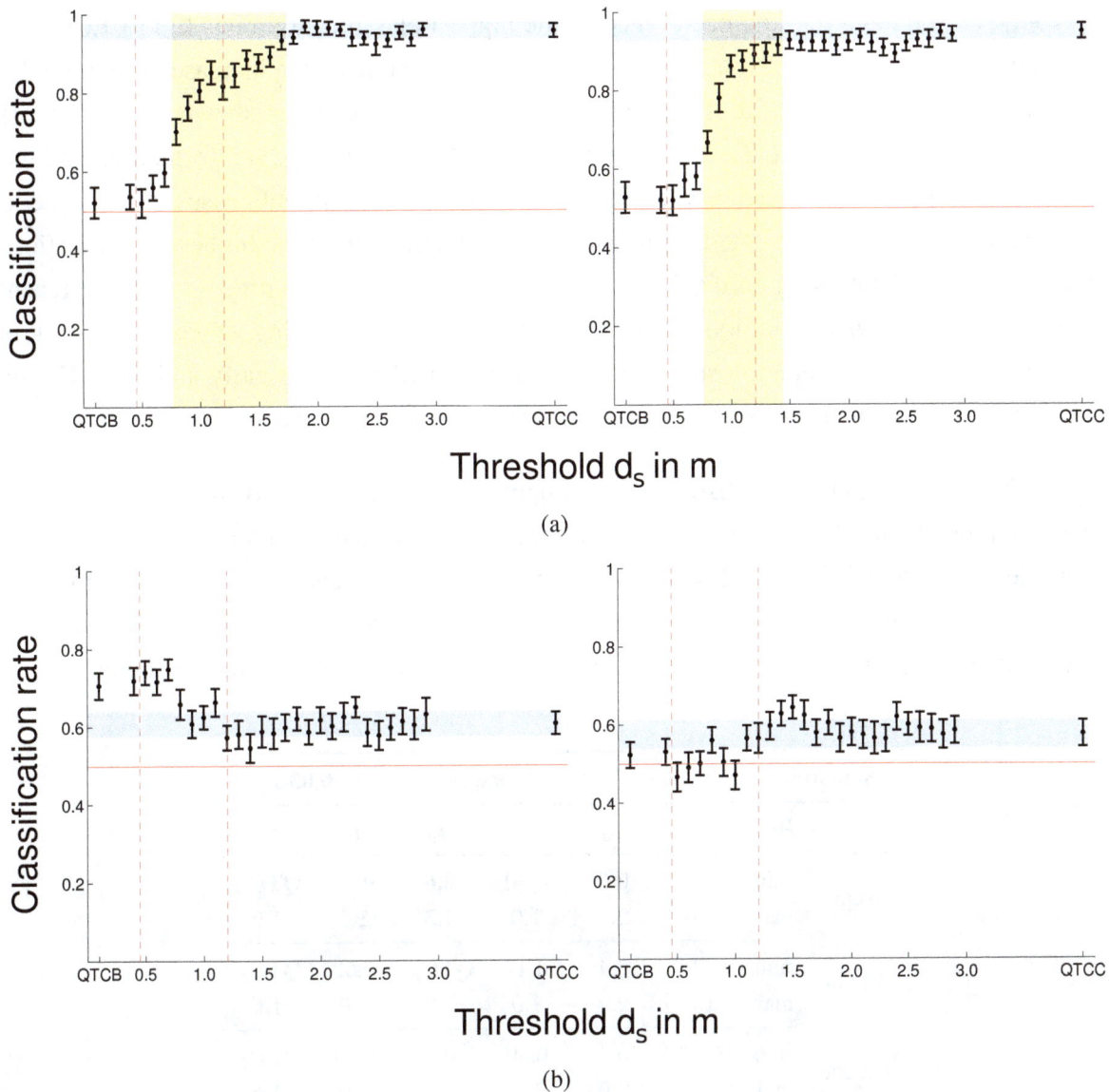

Figure 12. "Lincoln experiment" classification results. Dot: mean value, errorbar: 95% confidence interval, red line: H_0, left dashed red line: intimate space [8], right dashed red line: personal space [8]. The blue, horizontal area represents the 95% confidence interval of pure QTC_C for comparison. (**a**) Classification results for head-on passing on the left *vs.* right, lowest and highest smoothing parameters (see bold entries in Table 1b for min an max results). Left 1 cm and 0.1 s smoothing, right 10 cm and 0.3 s smoothing. The yellow, vertical area shows possible d_s where the left boundary represents the first distance d_s at which the two classes can be distinguished reliably and the right boundary shows the first value of d_s for which the classification results are not significantly different from QTC_C any more; (**b**) Classification results for head-on adaptive *vs.* non-adaptive. Left: 5 cm and 0.2 s smoothing, right: 1 cm and 0.3 s smoothing.

6.2. Results "Bristol Experiment"

Table 2 shows the evaluation of passing on the left *vs.* passing on the right using QTC_{BC} for the "Bristol experiment". The *early* condition, shown in Table 2a, shows its lowest classification rates for

$QTC_B \leq d_s \leq 0.6$ m, and the first occurrence of the highest classification rates (up to 1.0) for 1.6 m $\leq d_s \leq 2.3$ m. Reaching classification rates of 1.0 was made possible by the increase in training data for the "Bristol Experiment". Similar to the early condition, the *late* condition, shown in Table 2b, shows its lowest classification rates for $QTC_B \leq d_s \leq 0.6$ m, due to the missing 2D information, and the first occurrence of the highest classification rates for 1.5 m $\leq d_s \leq 2.4$ m. In both cases, 50% of the lowest classification rates have been generated using pure QTC_B, whereas all of the highest classification rates have been reached without using pure QTC_C. Classification rates of 1.0 with $p < 0.05$ are reached in 94% of the cases in the *early* condition and 100% in the *late* condition, using values of $d_s \geq 1.6$ m and $d_s \geq 1.5$ m respectively. Figure 13a shows the two unsmoothed cases for early and late. The *middle* condition is not shown here as it does not differ significantly from the two boundary cases.

Table 2. Classification results "Bristol Experiment": *Left vs. Right*, **bold:** *mentioned in text*. The mentioned confidence intervals represent the boundary cases and all the others can be considered lower. (**a**) Early. Maximum 95% confidence intervals ($p < 0.05$) for min and max classification results: *min:* 0.0333, *max:* 0.0066; (**b**) Late. Maximum 95% confidence intervals ($p < 0.05$) for min and max classification results: *min:* 0.0327, *max:* 0.0036.

(a)

Smoothing		**0.0 s**		**0.02 s**		**0.03 s**	
	Res.	μ	d_s	μ	d_s	μ	d_s
0 cm	min	0.49	0.6	0.50	**0.6**	0.52	**QTC$_B$**
	max	1.0	2.2	1.0	**2.3**	1.0	1.9
0.1 cm	min	0.48	0.4	0.47	**QTC$_B$**	0.52	**QTC$_B$**
	max	1.0	2.2	1.0	1.9	1.0	**1.6**
0.5 cm	min	0.47	0.4	0.50	0.4	0.54	**QTC$_B$**
	max	1.0	2.0	1.0	1.6	1.0	**1.6**
1 cm	min	0.58	0.4	0.47	**QTC$_B$**	0.52	**QTC$_B$**
	max	0.99	2.0	1.0	**1.6**	1.0	1.7

(b)

Smoothing		**0.0 s**		**0.02 s**		**0.03 s**	
	Res.	μ	d_s	μ	d_s	μ	d_s
0 cm	min	0.49	**QTC$_B$**	0.49	**QTC$_B$**	0.51	0.5
	max	1.0	2.3	1.0	**1.5**	1.0	1.6
0.1 cm	min	0.53	**QTC$_B$**	0.52	0.5	0.54	**0.6**
	max	1.0	2.3	1.0	**2.4**	1.0	1.6
0.5 cm	min	0.56	0.5	0.51	**QTC$_B$**	0.51	**QTC$_B$**
	max	1.0	2.0	1.0	2.0	1.0	1.6
1 cm	min	0.54	0.4	0.49	**QTC$_B$**	0.47	0.5
	max	1.0	2.0	1.0	**2.4**	1.0	1.6

(a)

(b)

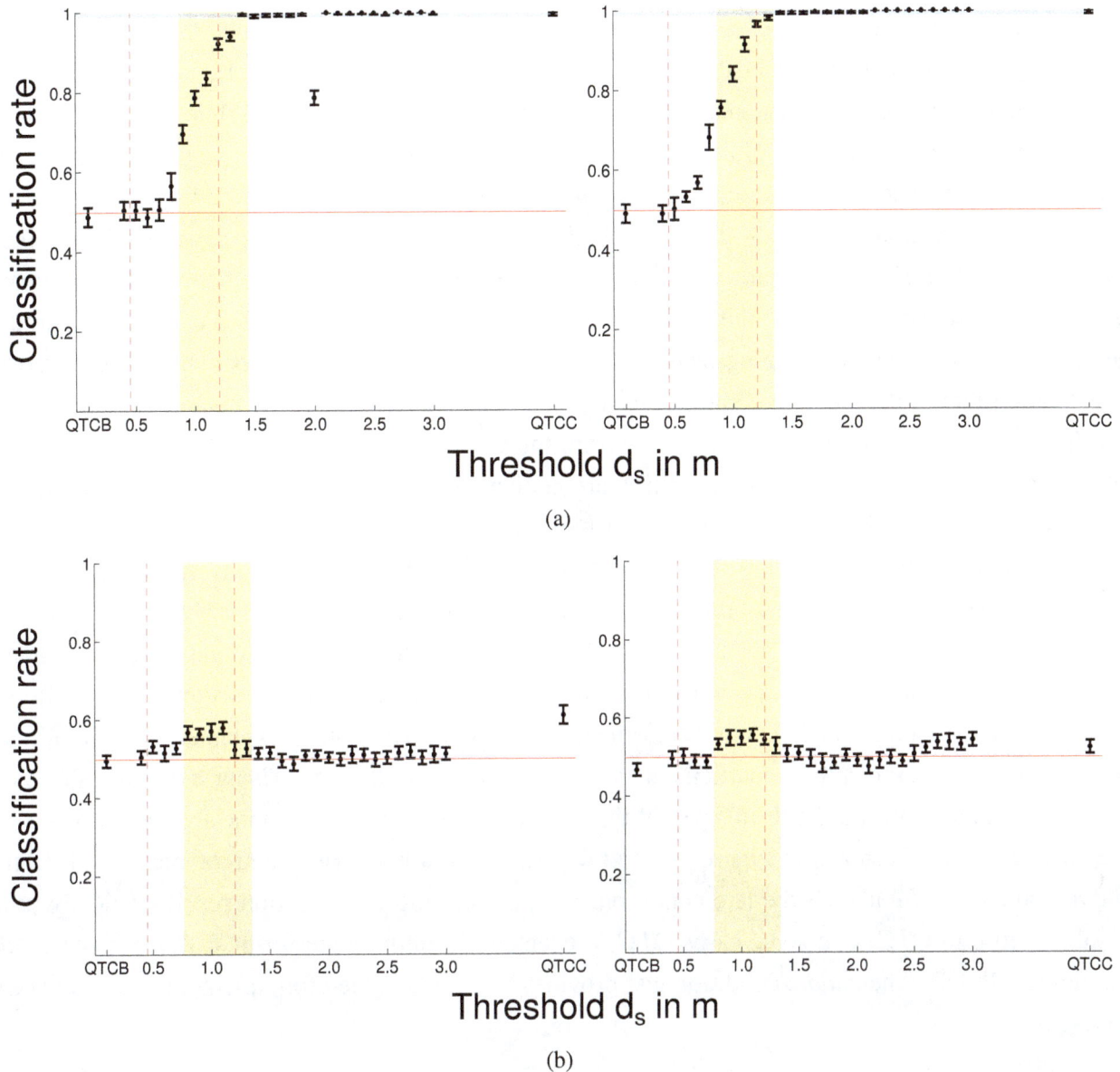

Figure 13. Classification results for *left vs. right* and *early vs. late*. Dot: mean value, errorbar: 95% confidence interval, red line: H_0, left dashed red line: intimate space [8], right dashed red line: personal space [8]. (**a**) Results for the *left vs. right* condition using unsmoothed data. *Left:* early condition, *right:* late condition. Significant classification results have been achieved for values $d_s > 0.8$ m regardless of the actual condition and reach optimal results for the classification using $d_s \approx 1.5$ m, see yellow, vertical area. The artefact at 2.1 m can be explained by the physical set-up of the experiment, *i.e.*, the corridor width. The increased confidence interval at 2.1 m is due to the "robot" getting tangled up in the curtains once. Blue, horizontal area: 95% confidence interval of pure QTC_C for comparison; (**b**) Results for the *early vs. late* condition. *Left:* unsmoothed data, *right:* highest smoothing values, *i.e.*, 1 cm and 0.03 s. Significant classification results have been achieved for values 0.8 m $\leq d_s \leq 1.3$ m regardless of the actual smoothing values, see yellow, vertical area. The good classification result for QTC_C with unsmoothed values might be due to artefacts from before the start or after the end of the interaction and must be very minute movements since they disappear when using even the lowest smoothing values.

Figure 13b shows the results for the comparison of the *early* and *late* condition. As can be seen form the figure, the two conditions can be distinguished for distances of 0.8 m $\leq d_s \leq$ 1.3 m, regardless of the actual smoothing values used. The majority of the values are not significantly different from H_0 except for the mention range of d_s. The influencing factor here is the actual minimum distances the participants kept to the experimenter in either condition. Fitting a normal distribution over the minimum distances kept in the early and late condition yielded a significant difference ($p < 0.05$): early: 0.98 m $\pm$ 0.02, late: 0.92 m $\pm$ 0.02, but the actual total difference between the mean values in the minimum distances for early and late is only 0.06 m; the slightly increased reaction time of $1.4s$ in the early compared to the late condition is the determining factor for this difference. Both these facts explain the improved classification rate in the mentioned range 0.8 m $\leq d_s \leq$ 1.3 m. As above, the *middle* condition is not shown because it does not significantly differ from the two other conditions. The minimum distances kept by the participant in the middle condition are neither significantly different from the early nor the late condition. Hence, classification cannot be achieved.

The results for the comparison of the *indicator vs. no indicator* conditions are very parameter dependent when it comes to smoothing and quantisation. Figure 14a shows the best result (left) and a typical result (right) for different smoothing and quantisation values in the late condition. The distance $d_s = 0.9$ m represents a special case where the classification rates jump to values significantly different from H_0 for all smoothing and quantisation value combinations. This can be explained by the minimum distance of 0.92 m to 0.98 m the participants kept to the robot at all times. Using a distance threshold of $d_s = 0.9$ m therefore highlights this part of the interaction by suppressing "unnecessary" information. The *early* condition is shown in Figure 14b and depicts the best result (left) and a typical result (right) in our evaluation. Similar to the late condition, at $d_s = 0.9$ the classification results typically jump to values close to QTC_C. In some cases QTC_{BC} even significantly outperforms QTC_C for certain d_s, see Figure 14b left. The *middle* condition just provides noise and is therefore unclassifiable via QTC_B, QTC_C, or QTC_{BC}.

(a)

(b)

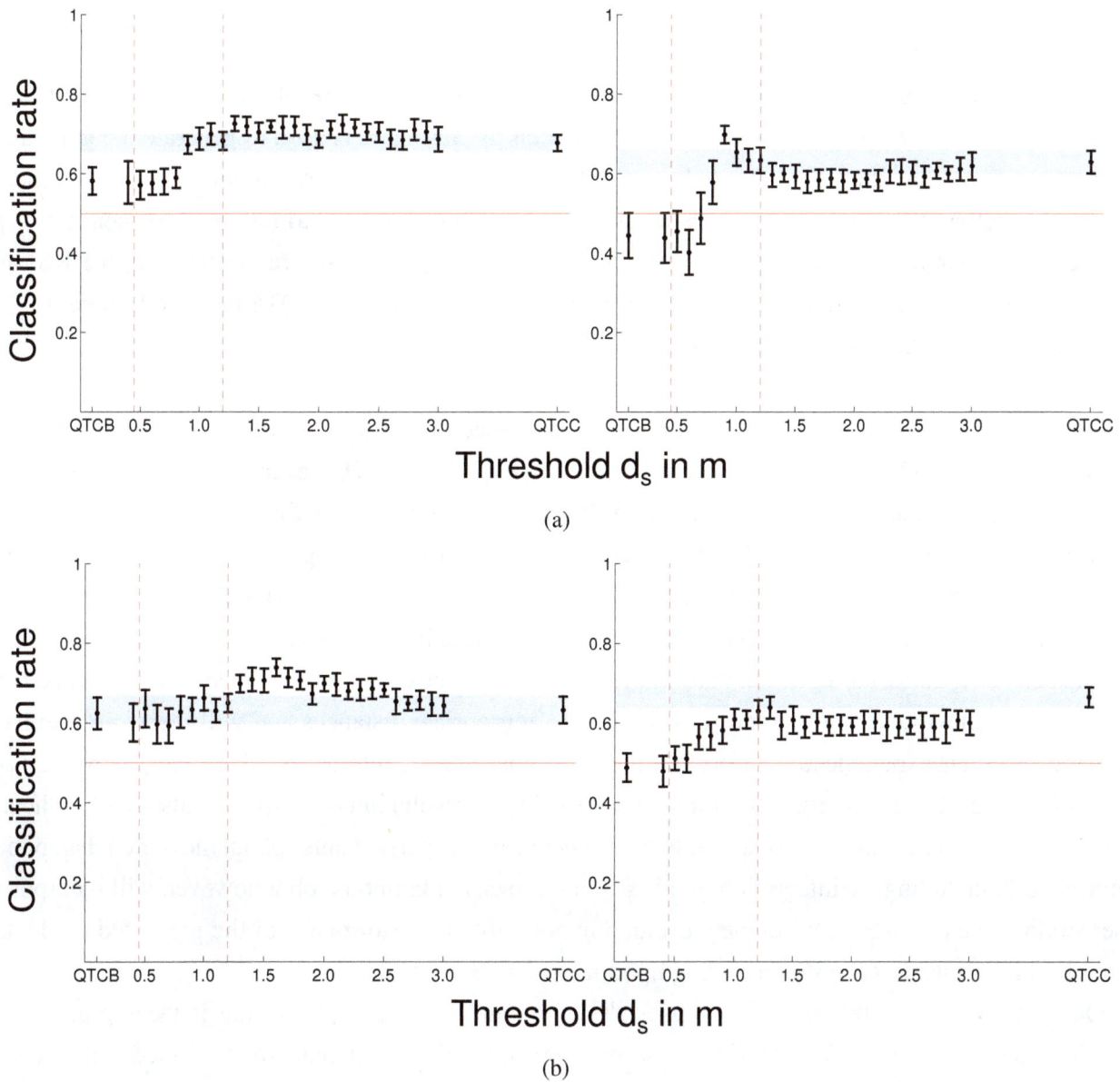

Figure 14. Classification results for *indicator vs. no indicator*. Dot: mean value, errorbar: 95% confidence interval, red line: H_0, left dashed red line: intimate space [8], right dashed red line: personal space [8]. Blue, horizontal area: 95% confidence interval of QTC_C showing that QTC_{BC} yields similar results for most values of d_s and significantly better results for certain distance values. (**a**) Results for *indicator vs. no indicator* in the late condition. *Left:* overall best results, smoothing values: 1 cm and 0.0 s, *right:* typical result, smoothing values: 1 cm and 0.02 s. The overall results are very dependent on the smoothing parameters. However, a significant jump in classification rates can be observed for $d_s = 0.9$ regardless of the actual smoothing values which can be explained by the model highlighting the distance at which the actual circumvention by the "robot" happened if there was any; (**b**) Results for *indicator vs. no indicator* in the early condition. *Left:* best result, smoothing values, 0.1 cm and 0.00 s, *right:* typical result, smoothing values 0.01 cm and 0.0 s. Results are very dependent on the smoothing parameters. Unsmoothed values contain too many artefacts to be useful for classification.

7. Discussion

In this section we focus on the interpretation of the classification results presented in Section 6. As described above, employing our probabilistic models as classifiers is used to generate a comparative measure to make assumptions about the quality of the generated representation where significant differences between the two used classes means that our model was able to reliably represent this type of interaction. We are evaluating the general quality of using QTC for the representation of HRSI and investigate the different distances or ranges of distances for the proposed QTC_{BC} based model to find suitable regions for the switch between the two variants.

Limitations   A possible limitation is that the presented computational model was not evaluated in a dedicated user study but on two data sets from previous experiments. However, a model of HRSI should be able to represent any encounter between a robot and a human in a confined shared space. The two used experiments might not have been explicitly designed to show the performance of the presented approach but provide the type of interactions usually encountered in corridor type situations, which represents a major part of human-aware navigation. Additionally, the instructions given in the "Bristol Experiment", to cross the corridor with as little veering as possible, might have also influenced the participants behaviour when it comes to keeping the appropriate distances and will therefore also have influence on their experienced comfort and the naturalness of the interaction. However, as we can see from Figure 15, the left *vs.* right conditions yielded similar results in both experiments which indicates that these instructions did not have a significant influence on the participants spatial movement behaviour. Future work, including the integration of this system into an autonomous robot however, will incorporate user studies in a real work environment evaluating not only the performance of the presented model but also of a larger integrated system building upon it.

Our presented probabilistic QTC_{BC} uses $d(k,l)_{t-1}$ and $d(k,l)_t$ to determine if the representation should transition from QTC_B to QTC_C or vice-versa. This might lead to unwanted behaviour if the distance $d(k,l)$ oscillates around d_s. This has to be overcome for "live" applications, e.g., by incorporating qualitative relations with learned transition probabilities for *close* and *far*. For the following discussion, due to the experimental set-up, we can assume that this had no negative effect on the presented data.

A general limitation of QTC is that actual sensor data does not coincide with the constraints of a continuous observation model represented by the CND. In the "Lincoln" data for example we encountered up to 521 illegal transitions which indicates that raw sensor data is not suitable to create QTC state sequences without post-processing. This however, was solved by using our proposed HMM based modelling adhering to the constraints defined in the CND, only producing valid state transitions.

A major limitation is that important HRSI concepts such as speed, acceleration, and distance, are hard to represent using QTC. While regular QTC_B is able to represent relative speeds, it is neither possible to represent the velocity nor acceleration of the robot or the human. Therefore, QTC alone is not very well suited to make statements about *comfort*, *naturalness*, and *sociability*, as defined by Kruse *et al.* [9], of a given HRSI encounter. We showed that, using implicit distance modelling is able to enrich QTC with such concepts but many more are missing.

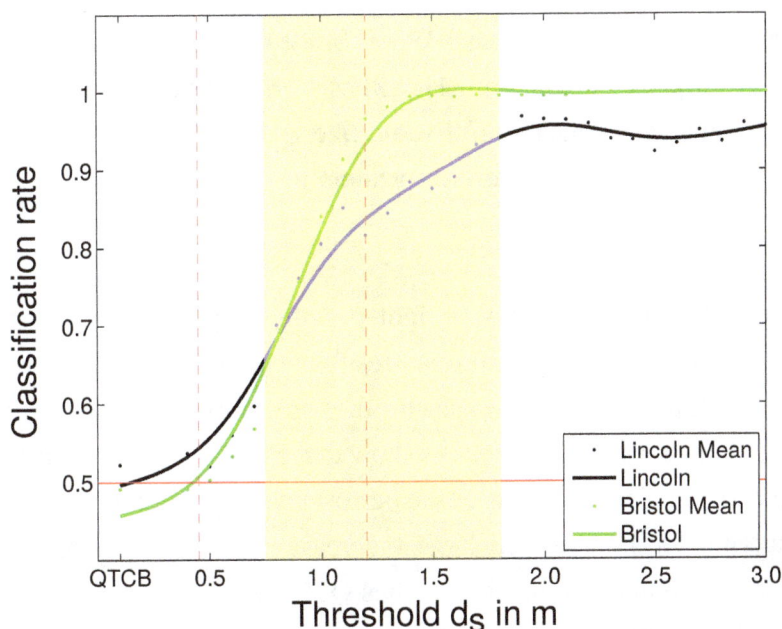

Figure 15. Comparing "Lincoln" and "Bristol" experiment results: Passing on the left *vs.* passing on the right. The blue curve represents the "Lincoln" experiment classification rates using the lowest smoothing values and the green curve represents the unsmoothed classification rate for the "Bristol" experiment, respectively. The curve has been obtained using a smoothing spline [47] with a *p*-value of $p = 0.99$. Red line: H_0, left dashed red line: intimate space [8], right dashed red line: personal space [8], yellow area: maximum interval for QTC_{BC} transitions. The better results for the "Bristol" experiment can be explained by the larger amount of training data.

Another limitation of QTC is the impossibility to infer which agent executes the actual circumvention action in the head-on scenario. When interpreting the graph in Figure 6, we are not sure if the human, the robot, or both are circumventing each other. We just know that the human started the action but we do not know if the robot participated or not. This could eventually be countered by using the full QTC_C approach including the relative angles. Even then, it might not be possible to make reliable statements about that and it would also complicate the graph and deprive it of some of its generalisation abilities.

Head-on *vs.* Overtake   The presented classification of head-on *vs.* overtaking (see Table 1a) shows that QTC_B, QTC_C, and QTC_{BC}, regardless of the chosen d_s, are able to reliably classify these two classes. We have seen that there are cases where pure QTC_B outperforms pure QTC_C. This is not surprising because the main difference of overtaking and head-on lies in the $(q_1\,q_2)$ 2-tuple of QTC_B, *i.e.*, both agents move in the same direction, e.g., $(-+)$, *vs.* both agents are approaching each other $(--)$. The 2D $(q_4\,q_5)$ QTC_C information can therefore be disregarded in most of the cases and only introduces additional noise. This indicates that QTC_B would be sufficient to classify head-on and overtaking scenarios but would of course not contain enough information to be used as a generative model or to analyse the interaction. QTC_{BC} allows to incorporate the information about which side robot and human should use to pass each other and the distance at which to start circumventing. Additionally, QTC_{BC} also allows to disregard information for interactants far apart, only employing the finer grained

QTC_C where necessary, *i.e.*, when close to each other. Since all of the found classification results were significantly different from $p = 0.5$–the Null Hypothesis (H_0) for a two class problem–this distance can be freely chosen to represent a meaningful value like Hall's personal space 1.22 m [8]. By doing so, we also create a more concise and therefore tractable model as mentioned in our requirements for HRSI modelling.

Left *vs*. Right The comparison of left *vs*. right pass-by actions in both experiments shows that using pure QTC_B does, not surprisingly, yield bad results because the most important information–on which side the robot and the human pass by each other–is completely omitted in this 1-dimensional representation. Hence, all the classification results show that an increase in information about the 2-tuple $(q_4\ q_5)$ representing the 2D movement increases the performance of the classification. On the other hand, the results of both experiments show that the largest increase in performance of the classifier happens at distances of $d_s \geq 0.7$ m and that classification reaches QTC_C quality at $d_s \geq 1.5$ m (see yellow area in Figures 12a, 13a and 15), which loosely resembles the area created by the far phase of Hall's personal space and the close phase of the social space [8]. These results could stem from the fact that the personal space was neither violate by the robot–be it fake or real–nor the participant. Judging from our data, the results indicate that information about the side $(q_4\ q_5)$ is most important if both agents enter, or are about to enter, each others personal spaces as can be seen from the yellow areas in Figures 12a, 13a and 15. The information before crossing this threshold can be disregarded and is not important for the reliable classification of these two behaviours. As mentioned in our requirements, recognising the intention of the other interactant is a very important factor in the analysis of HRSI. Reducing the information about the side constraint and only regarding it when close together, allows to focus on the part of the interaction where both agents influence each others paths and therefore facilitates intention recognition, based on spatial movement.

Figure 15 shows that our model gives consistent results over the two experiments in the left *vs*. right condition which is the only one we could compare in both. The blue curve shows the classification results for the "Lincoln experiment" whereas the green curve shows the results for the "Bristol experiment". Both curves show the same trends of significantly increasing classification results from 0.7 m $\leq d_s \leq 1.5$ m reaching their pinnacle at 1.5 m $\leq d_s \leq 2.0$ m. This implies that our model is valid for this type of interaction regardless of the actual environment set-up and that the fact that we used an autonomous robot in one of the experiments and a "fake robot" in the other does not influence the data. More importantly, it also shows a suitable distance range for this kind of HRSI that also encloses all the other found distance ranges from the other conditions and is therefore a suitable candidate for QTC_{BC} transitions.

Adaptive *vs*. Non-Adaptive Velocity Control Using a probabilistic model of pure QTC_C (as attempted in [13]), it was not possible to reliably distinguish between the two behaviours the robot showed during the "Lincoln experiment", *i.e.*, adaptive *vs*. non-adaptive velocity control. We investigated if QTC_{BC} would sufficiently highlight the part of the interaction that contains the most prominent difference between these two classes to enable a correct classification. Indeed, the results indicate that using a very low distance threshold d_s enables QTC_{BC} to distinguish between these two cases for some of the smoothing levels. In Figure 12b on the left side you can see the results from our previous work (using only pure QTC_C) [13] visualised by a horizontal blue area. The Figure also shows that some of the QTC_{BC} results are significantly different from QTC_C. Like for head-on *vs.* overtake, the main difference between the adaptive and non-adaptive behaviour seems to lie in the $(q_1\ q_2)$ 2-tuple, *i.e.*, both approach each other $(--)$ *vs.* human approaches and robot stops (-0). On the other hand, the classification rate drops to $p \approx 0.5$ (H_0) at $d_s = 1.3$ m most likely due to the increase in noise. Nevertheless, apart from these typical results, there is also an interesting example where this does not hold true and we see a slight increase in classification rate at $d_s = 1.5$ m which was the stopping distance of the robot (see Figure 12b, right). This shows that, even with QTC_{BC}, the results for adaptive *vs.* non-adaptive seem to be very dependent on the smoothing parameters (see Table 1c) and therefore this problem still cannot be considered solved. Incorporating another HRSI concept, *i.e.*, velocity or acceleration, might be able to support modelling of these kind of behaviours.

Early *vs.* Late   Looking at the data gathered in the "Bristol experiment", we also evaluated early *vs.* late (see Figure 13b) avoidance manoeuvres. Just to recapitulate, early means the "robot" executed the avoidance manoeuvre 700 ms before the indicator and in the late condition 700 ms after. The data shows that our model is able to represent this kind of interaction for distances of 0.8 m $\le d_s \le 1.3$ m. This is the distance the participants kept to the robot/experimenter in both experiments and loosely resembles Hall's personal space [8]. In this regard these results are consistent with the other described interactions showing that participants tried to protect their personal/intimate space. Except for the unsmoothed evaluation, we only achieved reliable classification using QTC_{BC} inside the mentioned range of 0.8 m $\le d_s \le 1.3$ m. QTC_B or QTC_C alone did not highlight the meaningful parts of the interaction and did not yield reliable results. Regarding the unsmoothed case, the fact that all the smoothing levels resulted in a significantly worse QTC_C classification than in the unsmoothed case shows that the unsmoothed result is most likely caused by artefacts due to minute movements before the start or after the end of the experiment. These movements cannot be regarded as important for the actual interaction and must therefore be considered unwanted noise.

Indicator *vs.* No Indicator   The "Bristol experiment" also used indicators (be it flashing lights or cartoon eyes) to highlight the side the "robot" would move to. In the control condition no indicators were used. Modelling these two conditions we can see from the late condition that for $d_s \ge 0.9$ m, which resembles the mean minimum distance kept by the participant, we can reliably distinguish the two cases. The classification rate does not improve significantly for greater distances or pure QTC_C but we are always able to reliably classify these two conditions. Compared to QTC_{BC} at $d_s \ge 0.9$ m, pure QTC_C shows worse results for some of the smoothing levels. This indicates that the most important part of the interaction happens at close distances (the mean minimum distance of both agents $d_s \approx 0.9$ m) and adding more information does not increase the accuracy of the representation or even decreases it.

8. Conclusions

In this work we presented a HMM-based probabilistic sequential representation of HRSI utilising QTC, investigated the possibility of incorporating distances like the concept of proxemics [8] into the model, and learned transitions in our combined QTC model and ranges of distances to trigger them, from real-world data. The data from our two experiments provides strong evidence regarding the generalisability and appropriateness of the representation, demonstrated by using it to classify different encounters observed in motion-capture data. We thereby created a tractable and concise representation that is general enough to abstract from metric space but rich enough to unambiguously model the observed spatial interactions between human and robot.

Using two different experiments, we have shown that, regardless of the modelled interaction type, our probabilistic sequential model using QTC is able to reliably classify most of the encounters. However, there are certain distances after which the "richer" 2D QTC_C encoding about the side constraint does not enhance the classification and thereby becomes irrelevant for the representation of the encounter. Hence, QTC_B's 1D distance constraint is sufficient to model these interactions when the agents are far apart. On the other hand, we have seen that there are distances at which information about the side constraint becomes crucial for the description of the interaction like in passing on the left *vs.* passing on the right. Thus, we found that there are intervals of distances between robot and human in which a switch to the 2-dimensional QTC_C model is necessary to represent HRSI encounters. These found distance intervals resemble the area of the far phase of Hall's personal space and the close phase of the social space, *i.e.*, 0.76 m to 2.1 m [8] (see Figure 15). Therefore, our data shows that using the full 2D representation of QTC_C is unnecessary when the agents are further apart than the close phase of the social space ($\approx$2.1 m) and can therefore be omitted. This not only creates a more compact representation but also highlights the interaction in close vicinity of the robot, modelling the essence of the interaction. Our results indicate that this QTC_{BC} model is a valid representation of HRSI encounters and reliably describes the real-world interactions in the presented experiments.

As a welcome side effect of modelling distance using QTC_{BC}, our results show that the quality of the created probabilistic model is, in some cases, even increased compared to pure QTC_B or QTC_C. Thereby, besides allowing the representation of distance and the reduction of noise, it also enhances the representational capabilities of the model for certain distance values and outperforms pure QTC_C. This shows the effect of reducing noise by filtering "unnecessary" information and focusing on the essence of the interaction.

Coming back to the four requirements to a model of HRSI stated in the introduction which were to

> *Represent the qualitative character of motions* to recognise intention, *represent the main concepts of HRSI* like proxemics [8], *be able to generalise* to facilitate knowledge transfer, and devise a *tractable, concise, and theoretically well-found model,*

we have shown that our sequential model utilising QTC_{BC} is able to achieve most of these. We exclusively implemented proxemics in our model which leaves room for improvement, incorporating other social norms, but shows that such a combination is indeed possible. Additionally, our representation is not only able to model QTC_B and QTC_C but also the proposed combination of both, *i.e.*, QTC_{BC}, which relies on the well founded original variants of the calculus. Therefore, the probabilistic sequential model based on QTC_{BC} allows to implicitly represent one of the main concepts of HRSI, distances. We do so by combining the different variants of the calculus, *i.e.*, the mentioned QTC_B and QTC_C, into one integrated model. The resulting representation is able to highlight the interaction when the agents are in close vicinity to one another, allowing to focus on the qualitative character of the movement and therefore facilitates intention recognition. By eliminating information about the side the agents are moving to when far apart, we also create a more concise and tractable representation. Moreover, the model also inherits all the generalisability a qualitative representation offers. This is a first step to employ learned qualitative representations of HRSI for the generation and analysis of appropriate robot behaviour.

Concluding from the above statements, the probabilistic model of QTC_{BC} is able to qualitatively model the observed interactions between two agents, abstracting from the metric 2D-space most other representations use, and implicitly incorporates the modelling of distance thresholds which, from the observations made in our experiments, represent one of the main social measures used in modern HRSI, proxemics [8].

9. Future Work

This research was undertaken as part of an ongoing project (the STRANDS project [48]) for which it will be used to generate appropriate HRSI behaviour based on previously observed encounters. The presented model will therefore be turned into a generative model to create behaviour as the basis for an online shaping framework for an autonomous robot. As part of this project, the robot is going to be deployed in an elder care home and an office building for up to 120 days continuously, able to collect data about the spatial movement of humans in real work environments.

To represent the interesting distance intervals we found in a more qualitative way, we will investigate possible qualitative spatial relations, e.g., *close* and *far*, ranging over the found interval of distances, instead of fixed thresholds. This will also be based on a probabilistic model using increasing transition probabilities depending on the current distance and the distance interval used. For example transitioning from QTC_B to QTC_C would be more likely the closer the actual distance is to the lower bound of the learned interval and vice-versa. In a generative system this will introduce variation and enables the learning of these distance thresholds.

To further improve this representation, we will work on a generalised version of our presented QTC_{BC} to deal with different and possibly multiple variants of QTC, which are not restricted to QTC_B and QTC_C, based also on other metrics beside Hall's social distances to allow behaviour analysis and generation according to multiple HRSI measures.

The experiments we used were originally meant to investigate different aspects of HRSI and not to evaluate our model explicitly. Some of the more interesting phenomena in the experiments, especially the "Bristol Experiment", like if the indicators had an effect on the interaction between the two agents or if the timing was important for the use of the indicators, will be investigated in more psychology focused work.

Acknowledgments

The research leading to these results has received funding from the European Community's Seventh Framework Programme under grant agreement No. 600623, STRANDS. The Bristol study was supported by an equipment grant from the Wellcome Trust (WT089367AIA) to set up the laboratory, and Brett Adey was supported by a Summer Vacation Scholarship, also by the Wellcome Trust. Alice Haynes was supported by an EPSRC Vacation Bursary. The authors would also like to thank Brett Adey and Alice Haynes for conducting the "Bristol experiment".

Author Contributions

Kerstin Eder and Ute Leonards conceived and designed the "Bristol" experiment; Christian Dondrup and Marc Hanheide conceived and designed the "Lincoln" experiment; Christian Dondrup performed the "Lincoln" experiment; Christian Dondrup analyzed the data of both experiments; Marc Hanheide and Nicola Bellotto contributed analysis tools; Marc Hanheide contributed the initial implementation of a Hidden Markov Model for the Qualitative Trajectory Calculus state chains; Nicola Bellotto contributed the correct definitions of QTC_B and QTC_C; Christian Dondrup contributed the definition of QTC_{BC}; Christian Dondrup contributed the implementation of the unified Hidden Markov Model of QTC_B, QTC_C, and QTC_{BC} state chains; Christian Dondrup wrote the paper.

Conflicts of Interest

The funding sponsors had no role in the design of the study; in the collection, analyses, or interpretation of data; in the writing of the manuscript, and in the decision to publish the results.

References

1. Steinfeld, A.; Fong, T.; Kaber, D.; Lewis, M.; Scholtz, J.; Schultz, A.C.; Goodrich, M. Common metrics for human-robot interaction. In Proceeding of the 1st ACM SIGCHI/SIGART Conference on Human-Robot Interaction-HRI '06, Salt Lake City, UT, USA, 2–3 March 2006; p. 33.

2. Borenstein, J.; Koren, Y. Real-time obstacle avoidance for fast mobile robots. *IEEE Trans. Syst. Man Cybern.* **1989**, *19*, 1179–1187.

3. Simmons, R. The curvature-velocity method for local obstacle avoidance. In Proceeding of the IEEE International Conference on Robotics and Automation, Minneapolis, MN, USA, 22–28 April 1996; Volume 4, pp. 3375–3382.

4. Sisbot, E.; Marin-Urias, L.; Alami, R.; Simeon, T. A Human Aware Mobile Robot Motion Planner. *IEEE Trans. Robot.* **2007**, *23*, 874–883.

5. Yoda, M.; Shiota, Y. Analysis of human avoidance motion for application to robot. In Proceeding of the 5th IEEE International Workshop on Robot and Human Communication, RO-MAN'96, Tsukuba, Japan, 11–14 November 1996; pp. 65–70.

6. Feil-Seifer, D.J.; Matarić, M.J. People-Aware Navigation For Goal-Oriented Behavior Involving a Human Partner. In Proceedings of the International Conference on Development and Learning, Frankfurt am Main, Germany, 24–27 August 2011.

7. Ducourant, T.; Vieilledent, S.; Kerlirzin, Y.; Berthoz, A. Timing and distance characteristics of interpersonal coordination during locomotion. *Neurosci. Lett.* **2005**, *389*, 6–11.

8. Hall, E.T. *The Hidden Dimension*; Anchor Books: New York, NY, USA, 1969.

9. Kruse, T.; Pandey, A.K.; Alami, R.; Kirsch, A. Human-aware robot navigation: A survey. *Robot. Auton. Syst.* **2013**, *61*, 1726–1743.

10. Peters, A. Small movements as communicational cues in HRI. In Proceeding of the 6th ACM SIGCHI/SIGART Conference on Human-Robot Interaction-HRI '11, Lausanne, Switzerland, 6–9 March 2011; pp. 72–73.

11. Hanheide, M.; Peters, A.; Bellotto, N. Analysis of human-robot Spatial behaviour applying a qualitative trajectory calculus. In Proceeding of the RO-MAN, Paris, France, 9–13 September 2012; pp. 689–694.

12. Bellotto, N.; Hanheide, M.; van de Weghe, N. Qualitative Design and Implementation of Human-Robot Spatial Interactions. In Proceeding of the International Conference on Social Robotics (ICSR), Bristol, UK, 27–29 October 2013.

13. Dondrup, C.; Bellotto, N.; Hanheide, M. A Probabilistic Model of Human-Robot Spatial Interaction using a Qualitative Trajectory Calculus. In Proceeding of the 2014 AAAI Spring Symposium Series, Palo Alto, CA, USA, 24–26 March 2014.

14. Dondrup, C.; Bellotto, N.; Hanheide, M. Social distance augmented qualitative trajectory calculus for Human-Robot Spatial Interaction. In Proceeding of the 23rd IEEE International Symposium on Robot and Human Interactive Communication, 2014 RO-MAN, Edinburgh, UK, 25–29 August 2014; pp. 519–524.

15. Van de Weghe, N. Representing and Reasoning about Moving Objects: A Qualitative Approach. Ph.D. Thesis, Ghent University, Ghent, Belgium, 2004.

16. Pacchierotti, E.; Christensen, H.I.; Jensfelt, P. Evaluation of Passing Distance for Social Robots. In Proceeding of the 15th IEEE International Symposium on Robot and Human Interactive Communication, RO-MAN 2006, Hafield, UK, 6–8 September 2006; pp. 315–320.

17. Lichtenthäler, C.; Peters, A.; Griffiths, S.; Kirsch, A. Social Navigation-Identifying Robot Navigation Patterns in a Path Crossing Scenario; In *Social Robotics*; Springer International Publishing, Switzerland, 2013; pp. 84–93.

18. Cohn, A.G.; Renz, J. Chapter 13 Qualitative Spatial Representation and Reasoning. In *Handbook of Knowledge Representation*; van Harmelen, F., Lifschitz, V., Porter, B., Eds.; Elsevier: Amsterdam, The Netherlands, 2008; Volume 3, pp. 551–596.

19. Torta, E.; Cuijpers, R.H.; Juola, J.F.; van der Pol, D. Design of robust robotic proxemic behaviour. In *Social Robotics*; Springer: Berlin Heidelberg, Germany, 2011; pp. 21–30.

20. Aldebaran Robotics. NAO robot: intelligent and friendly companion. Available online: https://www.aldebaran.com/en/humanoid-robot/nao-robot (accessed on 18 March 2015).

21. Walters, M.L.; Oskoei, M.A.; Syrdal, D.S.; Dautenhahn, K. A Long-Term Human-Robot Proxemic Study. In Proceeding of the 20th IEEE International Symposium on Robot and Human Interactive Communication, Atlanta, GA, USA, 31 July–3 August 2011; pp. 137–142.

22. Tranberg Hansen, S.; Svenstrup, M.; Andersen, H.J.; Bak, T. Adaptive human aware navigation based on motion pattern analysis. In Proceeding of the 18th IEEE International

Symposium on Robot and Human Interactive Communication, RO-MAN 2009, Toyama, Japan, 27 September–2 October 2009; pp. 927–932.

23. Kirby, R.; Simmons, R.; Forlizzi, J. COMPANION: A Constraint-Optimizing Method for Person-Acceptable Navigation. In Proceeding of the 18th IEEE International Symposium on Robot and Human Interactive Communication, Toyama, Japan, 27 September–2 October 2009; pp. 607–612.

24. Lu, D.V.; Allan, D.B.; Smart, W.D. Tuning Cost Functions for Social Navigation. In *Social Robotics*; Springer International Publishing: Switzerland, 2013; pp. 442–451.

25. Scandolo, L.; Fraichard, T. An anthropomorphic navigation scheme for dynamic scenarios. In Proceeding of the 2011 IEEE International Conference on Robotics and Automation (ICRA), Shanghai, China, 9–13 May 2011; pp. 809–814.

26. Svenstrup, M.; Bak, T.; Andersen, H.J. Trajectory planning for robots in dynamic human environments. In Proceeding of the 2010 IEEE/RSJ International Conference on Intelligent Robots and Systems (IROS), Taipei, Taiwan, 18–22 October 2010; pp. 4293–4298.

27. Martinez-Garcia, E.A.; Akihisa, O.; Yuta, S. Crowding and guiding groups of humans by teams of mobile robots. In Proceeding of the 2005 IEEE Workshop on Advanced Robotics and its Social Impacts, Nagoya, Japan, 12–15 June 2005; pp. 91–96.

28. Tamura, Y.; Dai Le, P.; Hitomi, K.; Chandrasiri, N.P.; Bando, T.; Yamashita, A.; Asama, H. Development of pedestrian behavior model taking account of intention. In Proceeding of the 2012 IEEE/RSJ International Conference on Intelligent Robots and Systems (IROS), Vilamoura, Algarve, Portugal, 7–12 October 2012; pp. 382–387.

29. Feil-Seifer, D.; Mataric, M. People-aware navigation for goal-oriented behavior involving a human partner. In Proceeding of the 2011 IEEE International Conference on Development and Learning (ICDL), Frankfurt am Main, Germany, 24–27 August 2011; Volume 2, pp. 1–6.

30. Garrido, J.; Yu, W. Trajectory generation in joint space using modified hidden Markov model. In Proceedingof the 23rd IEEE International Symposium on Robot and Human Interactive Communication, Edinburgh, UK, 25–29 August 2014; pp. 429–434.

31. Avrunin, E.; Simmons, R. Socially-appropriate approach paths using human data. In Proceeding of the 23rd IEEE International Symposium on Robot and Human Interactive Communication, Edinburgh, UK, 25–29 August 2014; pp. 1037–1042.

32. Kushleyev, A.; Likhachev, M. Time-bounded lattice for efficient planning in dynamic environments. In Proceeding of the IEEE International Conference on Robotics and Automation, ICRA'09, Kobe, Japan, 12–17 May 2009; pp. 1662–1668.

33. Tadokoro, S.; Hayashi, M.; Manabe, Y.; Nakami, Y.; Takamori, T. On motion planning of mobile robots which coexist and cooperate with human. In Proceeding of the 1995 IEEE/RSJ International Conference on Intelligent Robots and Systems 95. Human Robot Interaction and Cooperative Robots, Pittsburgh, PA, USA, 5–9 August 1995; Volume 2, pp. 518–523.

34. Ohki, T.; Nagatani, K.; Yoshida, K. Collision avoidance method for mobile robot considering motion and personal spaces of evacuees. In Proceeding of the 2010 IEEE/RSJ International Conference on Intelligent Robots and Systems (IROS), Taipei, Taiwan, 18–22 October 2010; pp. 1819–1824.

35. Bennewitz, M. Mobile robot navigation in dynamic environments. Ph.D. Thesis, University of Freiburg, Freiburg, Germany, 2004.

36. Bennewitz, M.; Burgard, W.; Cielniak, G.; Thrun, S. Learning motion patterns of people for compliant robot motion. *Int. J. Robot. Res.* **2005**, *24*, 31–48.

37. Ziebart, B.D.; Ratliff, N.; Gallagher, G.; Mertz, C.; Peterson, K.; Bagnell, J.A.; Hebert, M.; Dey, A.K.; Srinivasa, S. Planning-based prediction for pedestrians. In Proceeding of the IEEE/RSJ International Conference on Intelligent Robots and Systems, IROS 2009, St. Louis, MO, USA, 11–15 October 2009; pp. 3931–3936.

38. Chung, S.Y.; Huang, H.P. A mobile robot that understands pedestrian spatial behaviors. In Proceeding of the 2010 IEEE/RSJ International Conference on Intelligent Robots and Systems (IROS), Taipei, Taiwan, 18–22 October 2010; pp. 5861–5866.

39. Randell, D.A.; Cui, Z.; Cohn, A.G. A spatial logic based on regions and connection. In Proceedings of the 3rd International Conference on Principles of Knowledge Representation and Reasoning (KR'92), Cambridge, MA, 25–29, 1992; pp. 165–176.

40. Strands EU FP7 Project, No. 600623. Qualitative Spatial Representations Library. Available online: https://github.com/strands-project/strands_qsr_lib (accessed on 18 March 2015).

41. Delafontaine, M. Modelling and Analysing Moving Objects and Travelling Subjects: Bridging Theory and Practice. Ph.D. Thesis, Department of Geography, Ghent University, Ghent, Belgium, 2011.

42. Fink, G.A. *Markov Models for Pattern Recognition*; Springer-Verlag: Berlin Heidelberg, Germany, 2008.

43. Dondrup, C.; Lichtenthäler, C.; Hanheide, M. Hesitation signals in human-robot head-on encounters: A pilot study. In Proceedings of the 2014 ACM/IEEE International Conference on Human-Robot Interaction, ACM, Bielefeld, Germany, 3–6 March 2014; pp. 154–155.

44. Lincoln Centre for Autonomous Systems Research. Raw Study Data. Available online: https://github.com/LCAS/data (accessed on 18 March 2015).

45. Fox, D.; Burgard, W.; Thrun, S. The dynamic window approach to collision avoidance. *IEEE Robot. Autom. Mag.* **1997**, *4*, 23–33.

46. Thrun, S.; Burgard, W.; Fox, D. *Probabilistic Robotics*; MIT Press: Cambridge, MA, USA, 2005.

47. De Boor, C. *A Practical Guide to Splines*; Springer: New York, NY, USA, 1978.

48. Strands EU FP7 Project, No. 600623. The STRANDS Project. Available online: http://strands-project.eu (accessed on 18 March 2015).

7

Development of an Effective Docking System for Modular Mobile Self-Reconfigurable Robots Using Extended Kalman Filter and Particle Filter

Peter Won [1], Mohammad Biglarbegian [2,*] and William Melek [1]

[1] Mechanical and Mechatronics Engineering Department, University of Waterloo, Waterloo, ON, N2L 3G1, Canada; E-Mails: shown@uwaterloo.ca (P.W.); wmelek@uwaterloo.ca (W.M.)

[2] School of Engineering, University of Guelph, Guelph, ON, N1G 2W1, Canada

* Author to whom correspondence should be addressed; E-Mail: mbiglarb@uoguelph.ca

Academic Editor: Huosheng Hu

Abstract: This paper presents an autonomous docking system with novel integrated algorithms for mobile self-reconfigurable robots equipped with inexpensive sensors. A novel docking algorithm was developed to determine the initial distance and orientation of the two modules, and sensor models were established through experiments. Both Extended Kalman filter (EKF) and particle filter (PF) were deployed to fuse the measurements from IR and encoders and provide accurate estimates of orientation and distance. Simulation experiments were carried out first and then real experiments were conducted to verify the feasibility and good performance of the proposed docking algorithm and system. The proposed system provides a robust and reliable docking solution using low cost sensors.

Keywords: modular mobile self-reconfigurable robots; autonomous docking; state estimation; extended Kalman filter; particle filtering; IR sensor

1. Introduction

Modular and reconfigurable robots (MRR) consist of several modules enabling them to connect to each other to form various configurations. These robots can connect to each other autonomously to form a larger architecture that can perform tasks that one module cannot perform in terms of mobility, power, and functionality [1–4]. In order to autonomously connect two modules, the docking mechanism usually includes a mechanical system for connecting two modules and a sensing system that estimates the distance and orientations of the two modules.

Various docking mechanisms have been developed, including male-female, gripper, permanent magnet, and electromagnet. For successful autonomous docking, a distance or orientation sensing system is essential. An ultrasonic emitters and receivers have been used in determining the distance and orientation in [5,6]. Ultrasonic sensors are reliable and offer a wide range of measurement distances, but they require a line-of-sight [5]. Also, since they are vulnerable to sound noise, they cannot be used in noisy environments. Kim *et al.* in [7] estimated distance using RF and used a Kalman Filter (KF) for estimating the position and heading angle. However, authors concluded that extra sensors are required for this sensing system due to its inaccuracy. A camera is used to detect a target using scale-invariant feature transform (SIFT) method [8]. Although SIFT does not need any additional hardware, it has a high computational cost. In [9], a camera and LED lights are used to measure the distance and the orientation of the other module. A vision can also be used to differentiate modules based on color markers on each module [10]. A vision system with color markers does not require high computational power, but light condition and background object color may introduce errors.

Many reconfigurable robots use infrared (IR) emitter-receiver pairs to align their modules for docking due to its low-cost, small size, and low power consumption. The signal strength measurements depend on emitter angle, receiver angle, and the distance [11,12]. The signal strength is the maximum at a given distance when the IR emitter and receiver are aligned. In [13], this property of IR sensor was used to align two modules. However, due to the inaccuracy of the IR sensor, each module keeps rotating for alignment every time the module moves forward a short distance. In [14], the distance between two modules is measured using IR sensors while the orientation of each module is measured using magnetic sensors. However, magnetic sensors must be well calibrated and magnetic field distortion due to the presence of electric devices and ferromagnetic material is also a source of error.

Although accurate sensors are needed for successful docking, they are typically expensive and bulky. Therefore, development of an effective docking algorithm that uses low-cost small size sensors that can provide reliable data is desirable.

This paper presents a novel and robust autonomous docking system that uses EKF to estimate reliable distance and orientation data using low-cost IR emitters, IR receivers, and encoders. In most papers, IR sensors are used to estimate the distance assuming that emitter and receiver angles are close to zero, which often becomes a source of error. This paper presents IR sensor signal strength model based on distance, emitter angle, and receiver angle. Based on the IR sensor model and encoder measurements, an EKF is developed to accurately estimate the distance and angles of the two modules. In addition, an intelligent docking algorithm is developed to align two modules for a successful docking. A permanent magnet and electromagnet connector pair is proposed so that two modules can be docked successfully even with the presence of small offset and misalignment errors. The contributions of this paper lie in: (1) development of EKF and PF algorithms for estimating the position and orientation of autonomous modules; (2) development of a robust docking algorithm for mobile robots; and (3) experimental validation of the proposed system for different case studies, experiments were repeated several times to ensure robustness as well.

The rest of the paper is organized as follows: Section 2 describes the mechanical docking system as well as the sensing system including IR sensor modeling. Section 3 describes the proposed docking algorithm and the proposed EKF and PF algorithms. Section 4 discusses experimental results, while conclusions are provided in Section 5.

2. Docking System

This section discusses the proposed docking mechanism and the associated measurement system. For proper docking, sensors need to be accurately modeled; the sensory data will be used later in the docking algorithm. We first begin this section with introducing the docking mechanism. Next, we present models for the sensors used in this research. Figure 1 shows the proposed modular mobile self-reconfigurable robot. Each module is mobile and equipped with IR emitters and receivers, and encoders. As the docking mechanism, electromagnet and permanent magnet connectors are used.

Figure 1. Proposed modular mobile self-reconfigurable robot with sensing system.

2.1. Docking Mechanism

The reliability of the docking mechanism depends on several factors such as rigidity of the connections, error tolerance, and ease of connection/disconnection. To satisfy all the aforementioned factors, electromagnetic solution is considered the most effective mechanism. That is because the magnetic force self-aligns the magnets and also the connection between the modules is rigid. It should be noted that when two permanent magnets are used for docking, it is difficult to disconnect the two modules. Therefore, for easy disconnection, we use a permanent magnet for one side of a connector, and an electromagnet for the other side. This way the modules can be connected or disconnected.

Figure 2 shows our proposed electromagnet and permanent magnet connectors. On the first module (Module 1), a permanent magnet is connected to one side of a module via a long rod supported by a suction cup like a spring. Therefore, the suction cup allows the rod to move freely in two directions allowing the magnet to self-align for docking; the suction cup is also used to center the rod when it is not docked. A long rod is connected to a permanent magnet to compensate large offset errors when the modules are connecting. In order to center the magnet, a counterweight is connected at the other end of the rod. On the other module (Module 2), an electromagnet is connected to a universal joint so that the electromagnet can also move freely in two directions. When permanent magnet and electromagnet are close to each other, can be easily connected even if there is some misalignment and offset.

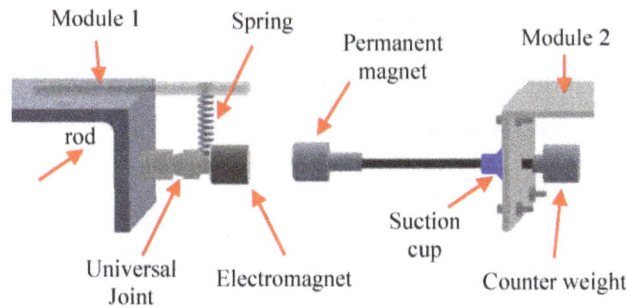

Figure 2. Proposed magnetic connector.

When the two modules are docked, the electromagnet is powered off, but the two modules stay connected because the permanent magnet attracts the electromagnet. To undock the modules, the polarity of the electromagnet is reversed so that it repels the permanent magnet. Therefore, the electromagnet is only powered during docking and undocking process resulting in small power consumption.

The proposed docking system could adjust up to 25 mm of offset error and 3° of misalignment error. These tolerances can be easily adjusted by changing the permanent magnet or input power to the electromagnet. The next subsection is devoted to the modeling of sensors; these models are used later in the docking algorithm.

2.2. IR Sensor Equation

The IR signal strength depends on the emitter and receiver angles as well as the distance to object. In [12] the IR signal strength was modeled as a function of the distance between emitter and receiver when the two are aligned, *i.e.*,

$$S(L) = \frac{a}{L^2} + \beta \tag{1}$$

where S is the signal strength, L is the distance between the emitter and the receiver, a is gain, and β is an offset. This offset includes ambient light and can be calculated by taking measurement when the emitter is off. The gain a can be found using a least-square fit method. In [15], both emitter and receiver are placed on the same side to measure the distance between the IR sensor and the object. The IR signal strength with respect to distance and angle was modeled as follows:

$$S(L, \theta) = \frac{a}{L^2} \cos\theta + \beta \tag{2}$$

where θ is the incident angle, and β is offset.

Equations (1) and (2) show that IR signal strength diminishes quadratically as the distance between the emitter and the receiver increases. Also, assuming that signal strength is a function of both emitter and receiver angles, the signal strength, S, can be modeled as [16]

$$S(L, \theta_e, \theta_r) = \frac{a}{L^2} \times f(\theta_e) \times f(\theta_r) + \beta \tag{3}$$

where $f(\theta_e)$ is a function of emitter angle and $f(\theta_r)$ is a function of receiver angle; emitter and receiver angles are illustrated in Figure 10a. For a successful docking, one needs to accurately measure the distance and orientation between emitter and receiver.

When the connectors are docked successfully, the distance between IR emitter and receiver become 12 cm due to the length of the rod and the universal joint (Figure 2).

Therefore, the IR sensor should have high sensitivity from 12 cm to 20 cm to accurately measure the distance and orientation for docking. To achieve a high sensitivity between 12 cm and 20 cm, the IR emitters should provide high radiant intensity. Also, it is desirable that the intensity measurement changes sharply as emitter or receiver angle changes. To satisfy these two conditions, a TSAL6200 emitter is chosen because it provides high radiant intensity and the angle of half intensity is only 17°. For the receiver, TCRT1000 is used. To model the signal strength as a function of three variables, two variables are kept constants and one variable is changed at a time. Also, the noise of the IR sensor is investigated to evaluate the reliability of the sensor and to incorporate it into the EKF.

2.2.1. Distance Equation

In this section, a model for the distance of IR sensor is developed. To do so, the IR measurements are taken at various distances while the emitter and the receiver are aligned (emitter and the receiver angles at 0°). Then, the signal strength becomes a function of only the distance between the emitter and the receiver. Figure 3 shows the signal strength measurements of the experiment at various distances and the developed distance model using a least square curve fitting.

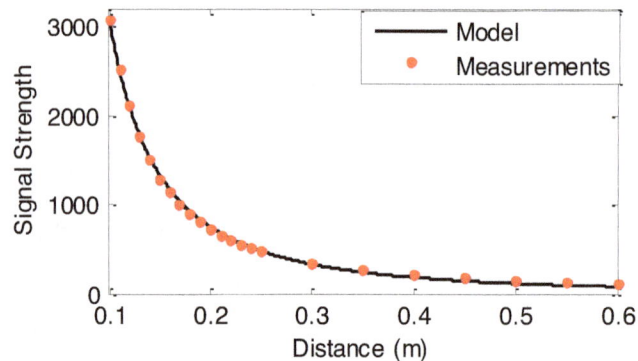

Figure 3. Signal strength measurements with respect to the distance between IR emitter and receiver and the distance equation when the emitter and receiver angles are 0°.

After compensating for the offset (β), our experimental results verify that the relationship between the distance and signal strength is:

$$S(L) = \frac{31.5}{L^2} \tag{4}$$

The signal strength is digital measurement (from 0 to 4095), thus does not have any unit. As can be seen in Figure 3, the equation matches the measurements very well.

2.2.2. Receiver Angle Equation

This section discusses the modeling of the receiver angle. For this purpose, the emitter angle is fixed at 0° and the distance is fixed at 15 cm and 25 cm. The measurements and the model with different distance are shown in Figures 4 and 5.

The distance equation was also incorporated into the equation of the signal strength as a function of the receiver angle. The relationship between the receiver angle (θ_r), distance (L), and signal strength (S) measurement is obtained as

$$S(L, \theta_r) = \frac{31.5}{L^2} \times \cos(1.12\theta_r) \tag{5}$$

where θ_r is the receiver angle (radian). Both Figures 4 and 5 show that the measurements match well with the equation given in Equation (5).

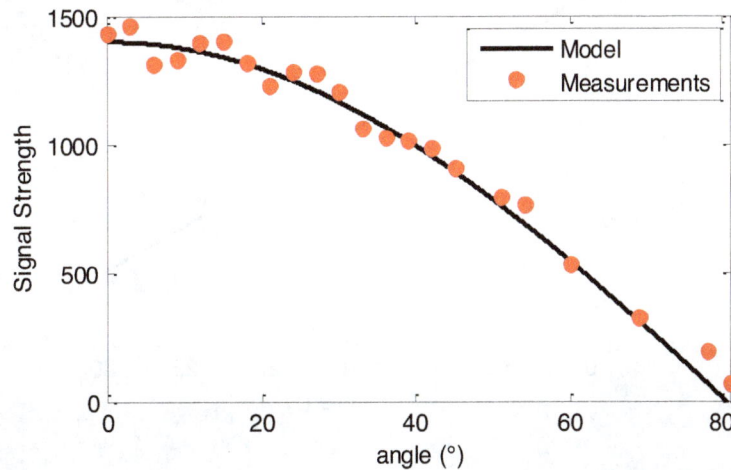

Figure 4. Signal strength with respect to the receiver angle measurements and the receiver at the distance of 15 cm.

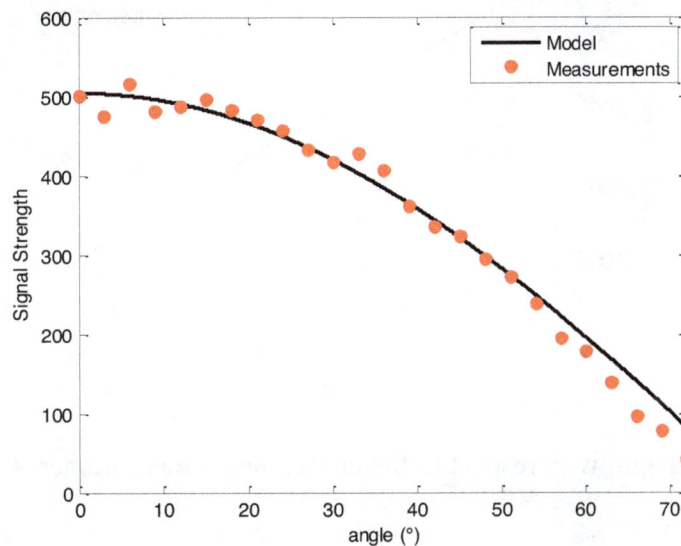

Figure 5. Signal strength with respect to the receiver angle measurements and the receiver at the distance of 25 cm.

2.2.3. Emitter Angle Equation

In this section, the signal strength is modeled with respect to the emitter angle. To do so, the receiver angle is fixed at 0° and the distance is fixed at 15 cm and 25 cm. The experiment results show that signal strength decreases linearly as the emitter angle increases. The relationship between the emitter angle (θ_e) and signal strength (S) measurement used for the model is obtained as

$$S(L, \theta_e) = \frac{47.7}{L^2}(0.66 - \theta_e) \tag{6}$$

where θ_e is emitter angle in radian. Since a linear equation with an offset is used, the estimated constant value is changed from 31.5 to 47.7, which is 31.5/0.66. The signal strength measurements and the model obtained are shown in Figures 6 and 7.

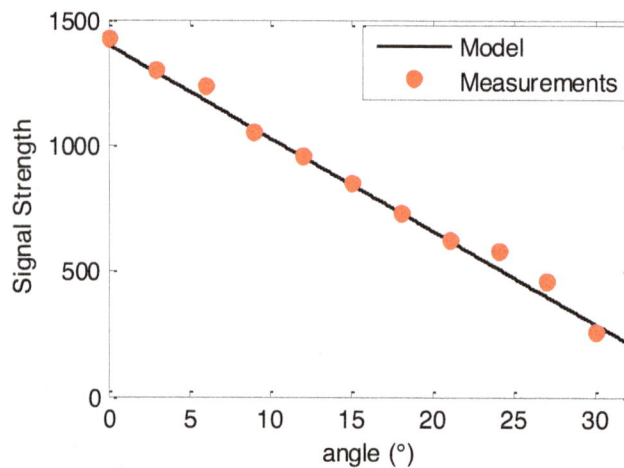

Figure 6. Signal strength with respect to the emitter angle measurements and the emitter model at the distance of 15 cm.

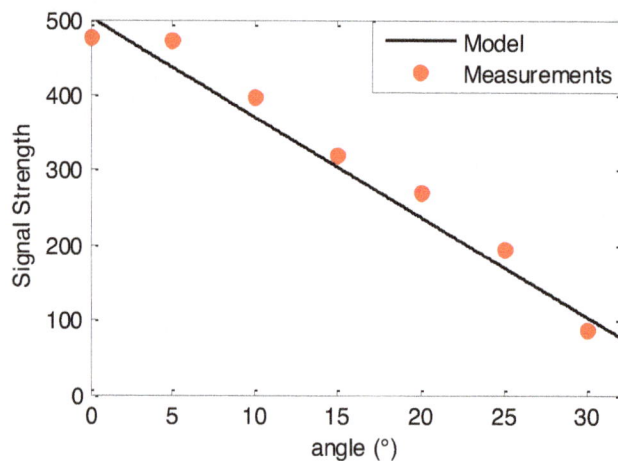

Figure 7. Signal strength with respect to the emitter angle measurements and the emitter at distance of 25 cm.

Equations (4)–(6) are now combined to model the signal strength as a function of distance, emitter angle, and receiver angle as:

$$S(L, \theta_r, \theta_e) = \frac{47.7}{L^2} \cos(1.12\theta_r)(0.66 - \theta_e) \tag{7}$$

2.2.4. Noise

We need to measure the IR sensor noise to evaluate the reliability of the sensor. We performed experiments where the results are shown in Figure 8. The measurements show that the noise increases almost linearly until signal strength is increased to 2500. Thus, the noise up to this region is modelled linearly. The noise measurements also show that when the signal is greater than 2500, the signal noise does not increase significantly and decreases when signal strength increases even further. For this application, the maximum signal strength measurement is 2200, which is measured when the IR emitter and receiver are 120 mm apart with 0° misalignment (Figure 3). Thus, we use a linear model for the noise.

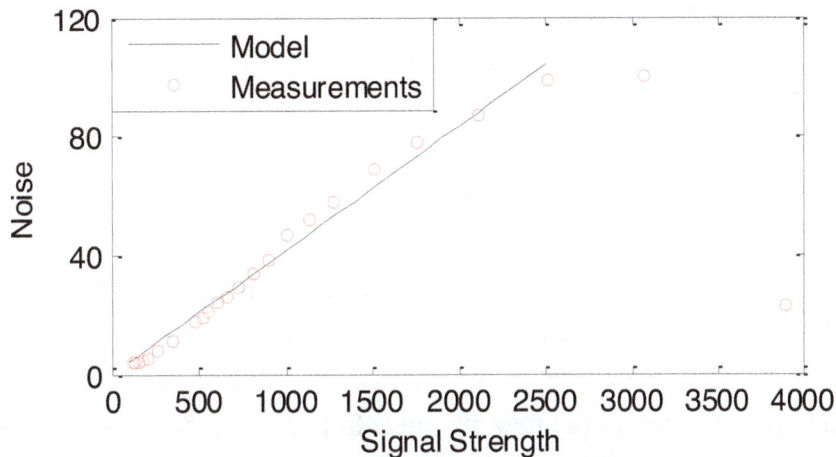

Figure 8. Root Mean Square (RMS) noise measurements.

3. Development of Docking Algorithm and State Estimators

In this section, we present a docking algorithm for two robot modules. To perform a successful docking our algorithm requires accessing the distance and angles of modules with respect to each other. Using "direct" sensory data is not possible because of inherent noise in sensors. We therefore, design estimators to estimate the required states (distance and orientation) for docking. Thus, we present our docking algorithm in Section 3.1. Next, we present the state estimators in Section 3.2.

3.1. Docking Algorithm

The proposed docking algorithm aligns two modules, determines the initial state, and check if the docking is successfully achieved (without using any extra sensors such as a limit switch). The docking algorithm is illustrated in Figure 9. The details of the algorithms are as follows:

1. When the docking is initiated, each module sends the IR signals to search for the other module.
2. Module 1 keeps rotating until it is approximately aligned with Module 2. To check for the alignment, the system detects if the signal strength decreases four consecutive times or it is 80% of the maximum measured signal strength. When Module 1 is approximately aligned, Module 1 rotates in the opposite

direction to compensate for the overshoot. Then, Module 1 stops and sends signals to Module 2 using the IR emitter to indicate its approximate alignment with Module 2.

3. For a fine alignment, IR measurements of Module 2 are used instead of Module 1 because signal strength decrease due to emitter angle changes (Figure 5) is much steeper compared to the receiver angle change (Figure 4) especially for small angles. When the IR signal strength of Module 2 decreases four consecutive times or the signal strength is 90% of the maximum measured signal strength, Module 2 signals Module 1 to stop rotating because Module 1 is now aligned with a small overshoot. Then, Module 1 rotates in the opposite direction for a short period of time to compensate for the overshoot.

4. After Module 1 is aligned, Module 2 is aligned using the same procedure (Step 2 to 3).

5. After modules are aligned, the distance between the two modules can be estimated using Equation (7).

6. Module 1 keeps moving forward. The state estimator is activated every time the encoder count exceeds a certain threshold.

7. If the emitter angle (θ_e) > 5° or IR measurement decreases, the initial estimation is most likely incorrect. In these cases, Module 1 stops moving forward and re-aligns. Then, the initial state is recalculated.

8. When the estimation of the emitter angle is 3° < θ_e < 5°, Module 1 adjusts the heading angle so that θ_e < 3° while moving forward.

9. If the distance, L, is less than 120 mm, which is the length of the docking part, Module 1 moves backwards for 3 seconds to check if docking is successful. If docking is successful, IR signal strength is high. Otherwise, IR signal strength is low because only Module 1 is moved backwards, which results in longer distance between the two modules. In this case, the docking procedure is repeated from Step 2.

State Estimators

Since the IR signal strength is a function of three different variables, it is difficult to accurately estimate distance between two modules especially from noisy measurements. Encoders can also be used to estimate the travelled distance and heading angle of a module, but it is not accurate due to wheels slippage, which accumulates the error over time especially for a mobile robot using a skid-steering system. To estimate distance and orientation more accurately from noisy IR sensor and inaccurate encoder measurements, an EKF and a PF are developed to estimate the required states. This section presents the development of the EKF and PF to estimate the distance and orientations of the modules using IR sensors and encoders. To use the state estimators more effectively, an efficient docking algorithm that aligns the two modules are proposed.

KF and EKF are widely used to estimate the position and orientation [17,18]. They combine multiple sensor measurements to achieve higher accuracy than using single sensor measurements. A KF is used to estimate states of linear models, and an EKF is used for nonlinear models. Since Equation (7) is nonlinear, an EKF should be used.

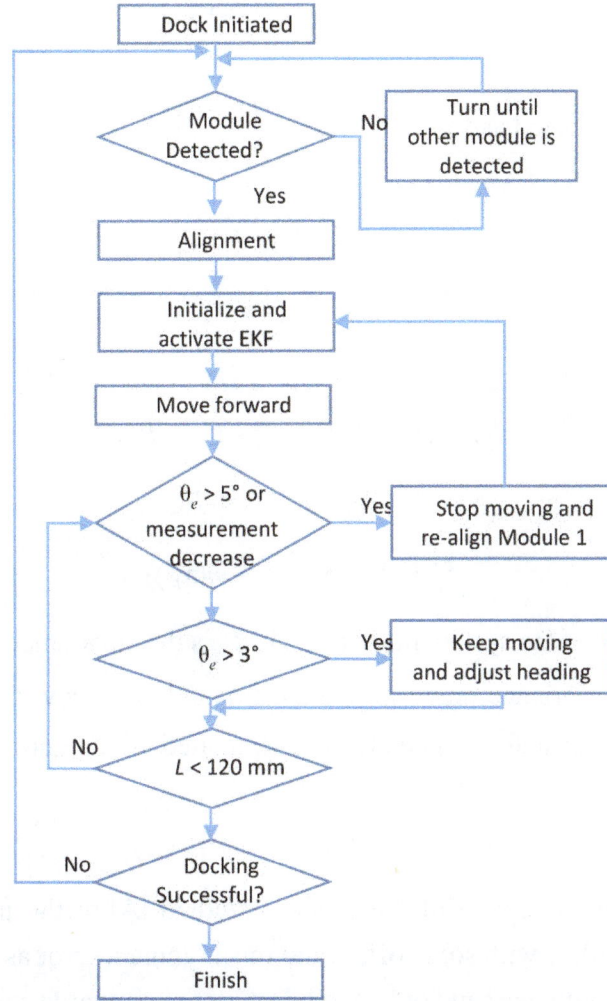

Figure 9. The Flow chart of the docking algorithm.

An EKF has two stages, prediction and update. Prediction stage predicts the next state using previous state estimate, input to the system, and the system model. Update stage corrects the prediction using measurements. Consider the following nonlinear state space model in a discrete time:

$$x_k = f_k\left(x_{k-1}, u_{k-1}\right) + w_{k-1} \tag{8}$$

$$z_k = h_k\left(x_k\right) + v_k \tag{9}$$

where subscript k represents iteration count, x_k is the state, f_k is a state transition function from time t_{k-1} to time t_k, u_{k-1} is the deterministic input, w_{k-1} is the process noise, z_k is the measurement, h_k is a measurement function, and v_k is the measurement noise. In order to use an EKF, the system models should be linearized using the first order Taylor series expansion. Let the first order Taylor series expansion of the nonlinear function $f_k\left(x_{k-1}, u_{k-1}\right)$ and $h_k\left(x_k\right)$ be A_k and C_k, respectively. Then, A_k and C_k become

$$A_k = f_k'\left(x_{k-1}, u_{k-1}\right) = \frac{\partial f_k\left(x_{k-1}, u_{k-1}\right)}{\partial x_{k-1}} \tag{10}$$

$$C_k = h'_k(x_k) = \frac{\partial h_k(x_k)}{\partial x_k} \tag{11}$$

Then, the state can be estimated as follows:
Prediction:

$$\hat{x}_k^- = f_k(\hat{x}_{k-1}^+, u_{k-1}) \tag{12}$$

$$P_k^- = A_k \cdot P_{k-1}^+ \cdot A_k^T + Q_{k-1} \tag{13}$$

Update:

$$K_k = P_k^- \cdot C_k^T \cdot [C_k \cdot P_k^- \cdot C_k^T + R_k]^{-1} \tag{14}$$

$$P_k^+ = [I - K_k \cdot C_k] \cdot P_k^- \tag{15}$$

$$\hat{x}_k^+ = \hat{x}_k^- + K_k \cdot (z_k - h(\hat{x}_k^-)) \tag{16}$$

where Q_{k-1} is the covariance of the system noise (w_{k-1}), R_k is the covariance of measurement noise (v_k), P_k^- is the predicted error covariance, and P_k^+ is the estimated error covariance after measurement. In order to use an EKF, the initial state (x_0) needs to be estimated with a good accuracy.

3.2. EKF Model

To develop an EKF, the movement of the robot is studied. After the initial alignment, Module 2 should face the rear of Module 1 with some offset and misalignment error as shown in Figure 10a. After the alignment, Module 2 is stationary and only Module 1 moves forwards and rotates for docking. Thus, the positions of emitter (E2) and receiver (R2) do not change. From Figure 10a, the relationship among the Module 1 heading direction (θv), emitter angle (θe), and receiver angle (θr) is

$$\theta e = |\theta v + \theta r| \tag{17}$$

The emitter angle and receiver angle cannot be negative because the maximum IR strength is at $0°$, and whether the angle increase clockwise or counter clockwise, IR strength measurement decreases That is why emitter and receiver angles should be absolute values.

From Figure 10b, when Module 2 moves forwards, current distance, L_k, with the heading angle at θv changes by ΔL_k, resulting in a new distance L_{k+1}. Therefore, the distances at step $k + 1$ can be estimated using a cosine law as follows:

$$L_{k+1}^2 = L_k^2 + \Delta L_k^2 - 2 \times L_k \times \Delta L_k \times \cos\theta r_k \tag{18}$$

Under the assumption that θv is kept constant. For this application, θv can be assumed constant because the module heading angle change ($\Delta\theta v$) is small for each KF iteration step.

The distance change of Equation (19) is calculated as [19]:

$$\Delta L_k = \frac{\Delta DR_k + \Delta DL_k}{2} \tag{19}$$

where ΔDR_k and ΔDL_k are the travelled distance of the right wheel and the left wheel during the time step, respectively, which are used as deterministic input, u_k, in Equation (12).

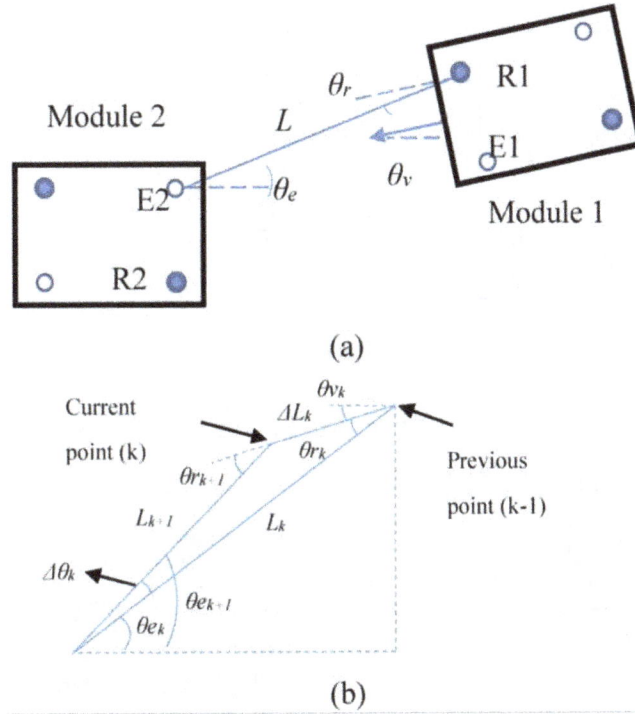

(a)

(b)

Figure 10. Schematics of the distances between IR sensors, and emitter and receiver angles. R1 and R2 represent receivers and E1 and E2 represent emitters. (**a**) Schematic of two modules and (**b**) relationship between receiver 1 (R1) and emitter 2 (E2) at time k.

The heading angle change due to the rotation of Module 1 can be calculated as:

$$\theta v_{k+1} = \theta v_k + \Delta\theta v_k = \theta v_k + \frac{\Delta DR_k - \Delta DL_k}{2w} \tag{20}$$

where w is the width of the vehicle (length between the two wheels). The orientation change ($\Delta\theta v_k$) calculation is shown in [19]. After Module 2 moves by ΔL_k, the emitter angle and the receiver angle are changed to

$$\theta e_{k+1} = \theta e_k + \Delta\theta_k \tag{21}$$

$$\theta r_{k+1} = \theta r_k + \Delta\theta_k \tag{22}$$

Where

$$\Delta\theta_k = acos\left(\frac{L_k^2 + d^2 - \Delta L_k^2}{2d \times L_k}\right) = acos\left(\frac{2L_k^2 - 2L_k\Delta L_k \cos\theta r_k}{2d \times L_k}\right)$$

$$= acos\left(\frac{L_k - \Delta L_k \cos\theta r_k}{d}\right) \tag{23}$$

Distance squared, heading angle and receiver angle are used as the state variables as:

$$x_k = [L_k^2 \ \theta v_k \ \theta r_k]^T \tag{24}$$

Then, the system matrix, A_k, can be expressed as

$$A_k = \begin{bmatrix} \dfrac{\partial f_{L^2}}{\partial L^2} & \dfrac{\partial f_{L^2}}{\partial \theta_v} & \dfrac{\partial f_{L^2}}{\partial \theta r} \\[2mm] \dfrac{\partial f_{\theta v}}{\partial L^2} & \dfrac{\partial f_{\theta v}}{\partial \theta_v} & \dfrac{\partial f_{\theta v}}{\partial \theta r} \\[2mm] \dfrac{\partial f_{\theta r}}{\partial L^2} & \dfrac{\partial f_{\theta r}}{\partial \theta_v} & \dfrac{\partial f_{\theta r}}{\partial \theta r} \end{bmatrix}_{x_k} = \begin{bmatrix} 1 - \dfrac{\Delta L_k \times \cos \theta r_k}{L_k} & 0 & 2 \times L_k \times \Delta L_k \times \sin \theta r_k \\[2mm] 0 & 1 & 0 \\[2mm] \dfrac{-\Delta L_k \times \sin \theta r_k}{2 L_k \times d^2} & 0 & 1 \end{bmatrix}_{x_k} \tag{25}$$

where

$$d^2 = L_k^{\,2} + \Delta L_k^{\,2} - 2 \times L_k \times \Delta L_k \times \cos \theta r_k \tag{26}$$

For the measurement model, $h_k(x_k)$, Equation (7) is used. In a general form, the signal strength is given by

$$\begin{aligned} S_k &= \frac{a}{L_{k+1}^{\,2}} \times \cos(c \times \theta r_k) \times (b - \theta e_k) \\ &= \frac{a}{L_{k+1}^{\,2}} \times \cos(c \times \theta r_k) \times (b - |\theta r_k + \theta v_k|) \end{aligned} \tag{27}$$

Then, the linearized measurement matrix can be determined from Equation (27) as

$$C_k = \begin{bmatrix} \dfrac{-a \times (b - \mathrm{abs}(\theta r_k + \theta v_k)) \times \cos(c \times \theta r_k)}{L_{k+1}^{\,4}} & -\dfrac{a \times \cos(c \times \theta r_k)}{L_{k+1}^{\,2}} \\[4mm] -\dfrac{a \times \cos(c \times \theta r_k)}{L_{k+1}^{\,2}} - \dfrac{a \times c \times (b - |\theta r_k + \theta v_k|) \times \sin(c \times \theta r_k)}{L_{k+1}^{\,2}} \end{bmatrix}$$

when $(\theta r_k + \theta v_k) > 0$ and

$$C_k = \begin{bmatrix} \dfrac{-a \times (b - \mathrm{abs}(\theta r_k + \theta v_k)) \times \cos(c \times \theta r_k)}{L_{k+1}^{\,4}} & -\dfrac{a \times \cos(c \times \theta r_k)}{L_{k+1}^{\,2}} \dfrac{a \times \cos(c \times \theta r_k)}{L_{k+1}^{\,2}} \\[4mm] -\dfrac{a \times c \times (b - |\theta r_k + \theta v_k|) \times \sin(c \times \theta r_k)}{L_{k+1}^{\,2}} \end{bmatrix} \tag{28}$$

when $(\theta r_k + \theta v_k) < 0$.

3.3. Particle Filter

PF [20,21] approximates posterior using state samples, called particles, with the corresponding weights. First, the PF places a finite number of states (particles) based on the previous posterior. Then, the next state of the particle is predicted using the dynamic model. Using the predicted states and the measurements, the weight of each particle is determined as:

$$w_k^i \propto p(z_k \mid x_k^i) \tag{29}$$

where x_k^i represents i^{th} particle of kth iteration and w_k^i represents the corresponding weight of particle x_k^i.

When the weights of all particles are determined, they are normalized. Then, the state is estimated based on the states and their corresponding weights as:

$$x_k = \sum w_k^i \times x_k^i \tag{30}$$

Then, new particles are resampled based on the weights and the covariance of the system noise.

To achieve high accuracy, the number of particles should be large enough to represent the posterior distribution. Although PF does not require calculating error covariance, Kalman gain, and inverse of matrices, it requires higher computational power than an EKF due to the requirement of a large number of particles for accurate state estimation.

For the proposed PF, the same states as EKF are used as shown in Equation (24). For the initial state distribution, the receiver angle and heading angle of the particles are spread evenly, and the distance of each particle is calculated based on measurement, receiver angle and heading angle using Equation (7).

For the prediction model, Equations (18), (20) and (22) are used, and for the measurement model, Equation (27) is used.

4. Simulation and Experiments

In this section, we provide simulation and experimental results to validate the performance of developed estimators. In addition, in the experiments, we demonstrate the docking performance for two mobile modules and repeat the experiments several times to prove its robustness as well.

4.1. Simulations

Both EKF and PF were simulated to determine which state estimator offers a better solution for this application. For the simulations of PF, the initial heading angle and receiver is spread from −0.05 radian to 0.05 radian (2.9°). Our PF uses 121 particles, because after several simulations we found out increasing the number of particles beyond 121 will not improve the results considerably and instead will add to the computational cost. Therefore, for the PF we keep the number of particles to be 121, which is optimum, both in terms of accuracy and computational cost.

Two different case studies are considered and the results of EKF and PF were compared. These two case studies are: 1) when the initial estimations are accurate, and 2) when initial estimations are not accurate.

It was assumed that the encoders have standard deviation errors of 10% due to slippage of the robot tires, and the IR sensor has a standard deviation error of 4% of the IR signal strength.

4.1.1. Using Correct Initial Position and Orientations

The initial position and orientations of the two modules are accurately estimated for the first simulation. The emitter and the receiver are 270 mm apart, and the two modules are perfectly aligned, meaning both heading angle and receiver angle are 0°. The positions of Module 1 at different time steps of simulation are shown in Figure 11. The green circle represents true position of the Module 1, the black rectangle represents position estimation using the EKF, red "x" marks represents particles of the PF, and "+" represents the final states which is the average of the PF states.

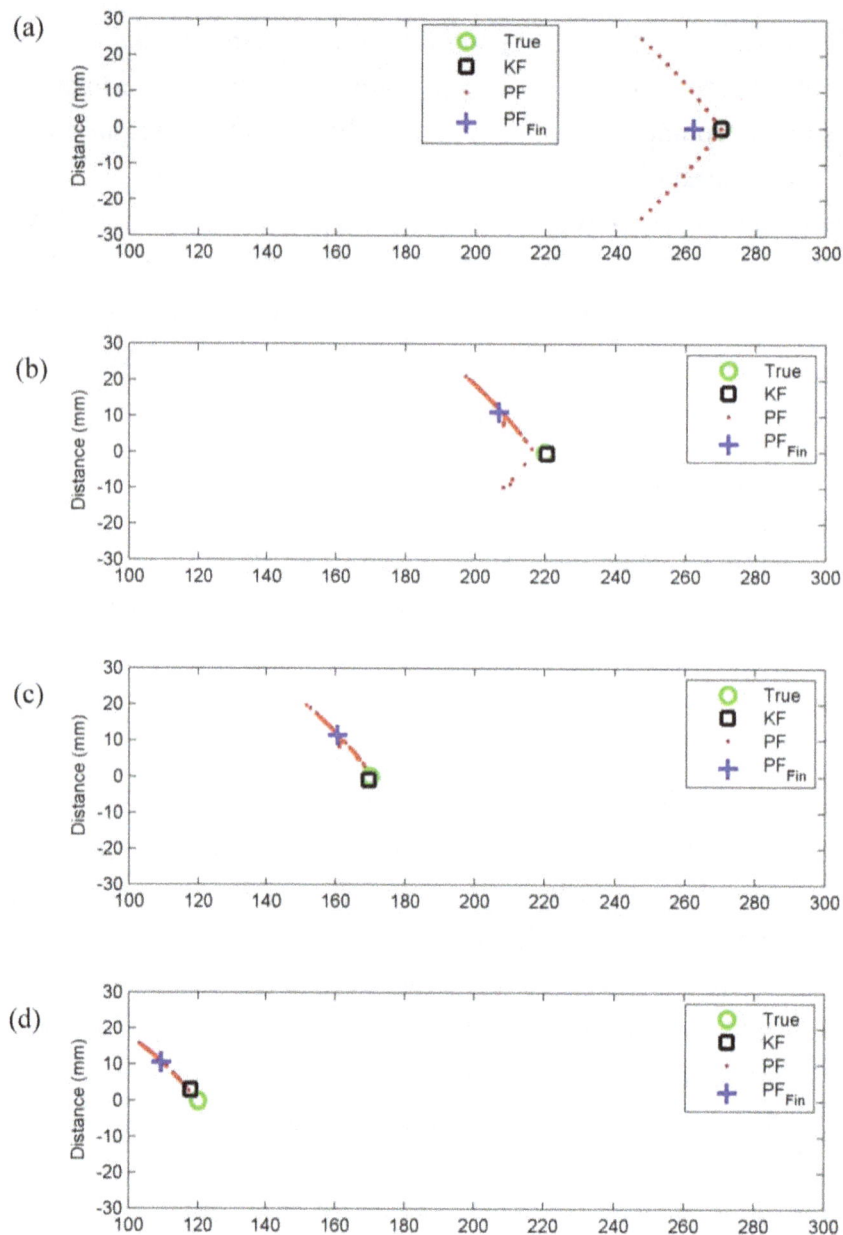

Figure 11. Position comparison of truth, EKF, and PF when the initial estimation is accurate. (**a**) initial position, (**b**) after moving 50 mm, (**c**) after moving 100 mm, and (**d**) after moving 150 mm, which represents docking position

Figure 11a shows that the EKF position matches the true position because the initial estimation is known. The particles of PF are spread evenly from 0° as well. Figure 11b shows the states after Module 1 moves 50 mm forward. It shows that particles are more condensed than Figure 11a because all the particles with high errors are eliminated. The EKF position estimation has some error compared to the true value. As iteration continues, particles get closer to the true value. On the other hand, the EKF error tends to increase as time proceeds, but shows better accuracy than PF. The state estimations of this simulation are shown in Table 1. The table shows that estimations of EKF are more accurate than PF.

Table 1. Position comparison when initial estimation of EKF is correct.

	Position (mm)	Heading Angle (°)	Receiver Angle (°)
True	120.4	0	0
EKF	117.8	0.83	0.66
PF	110.0	1.78	3.89

In order to quantify the error more accurately, the simulation is repeated 200 times and the average error of EKF and PF are evaluated (Table 2). Table 2 demonstrates that the EKF has much lower error than the PF for this application.

Table 2. Average state estimation error with 200 simulations.

	Position (mm)	Position δ	Heading angle (°)	Heading δ	Emitter angle (°)	Emitter δ
EKF	2.2	1.7	1.10	1.40	0.57	0.74
PF	8.7	5.0	1.03	0.93	4.32	2.82

4.1.2. Using Incorrect Initial Position and Orientations

The simulation was repeated with the same condition as previous simulation but the true heading angle of Module 1 is set to 0.05 radian (2.86°) and emitter angle is set to 0.05 radian. The simulation results are shown in Figure 12. Figure 12a shows that the EKF position does not match the true position due to the initial estimation error. Since it was assumed that the two modules are aligned, Figure 12a shows that the initial states of EKF is far behind the true position. Although the iteration continues, EKF position estimation does not converge to the correct position due to the incorrect initial estimation. On the other hand, PF converges to the true estimation as the iteration continues.

Table 3 shows the comparison between PF, EKF, and true state estimations after two modules are docked. Table 3 shows that the PF estimation converges to the correct states, but EKF estimation does not.

The simulations were repeated 200 times to study the average error of EKF and PF as shown in Table 4. Estimation errors of PF of two different simulations (Tables 2 and 4) are not significantly different. However, the state estimation error of EKF increased significantly when the initial estimation error is not accurate.

The simulations show that EKF is more accurate when the initial estimation is known, but when the initial estimation is not accurately estimated, EKF does not converge to the correct value. However, PF converges to the correct state because the fittest particles survive. It should be noted that the EKF iteration was 130 times faster than the PF iteration with 121 particles in this simulation. Also, even though it is not shown on this simulation, the estimation error of PF increases if the initial error is far away from the initial samples of PF (*i.e.*, receiver angle = 0.15 rad). Thus, EKF is considered as a better solution if the initial state is correctly estimated especially for an application that uses autonomous robots with limited on-board CPU computational power.

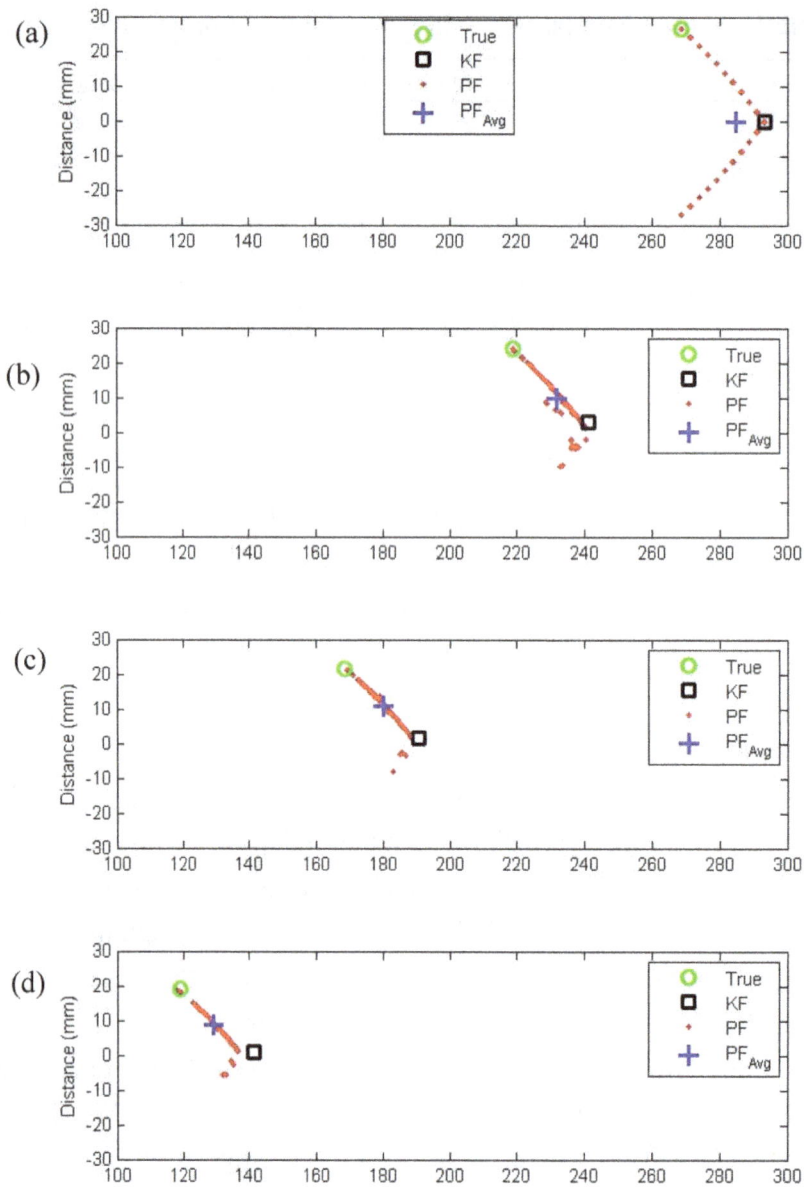

Figure 12. Position comparison of truth, EKF, and PF when the initial estimation is not correct. (**a**) initial position, (**b**) after moving 50 mm, (**c**) after moving 100 mm, and (**d**) after moving 150 mm, which represents docking position

Table 3. Position comparison when initial estimation of EKF is not correct.

	Position (mm)	Heading Angle (°)	Receiver Angle (°)
True	120.4	2.86	6.43
EKF	141.5	0.42	0.01
PF	129.7	0.85	3.18

Table 4. Average state estimation error with 200 simulations.

	Position (mm)	Position δ	Heading angle (°)	Heading δ	Emitter angle (°)	Emitter δ
EKF	23.1	3.0	4.50	1.57	6.22	0.39
PF	9.9	5.4	3.16	1.09	2.62	2.40

4.2. Experiments

Two modules were built and docking experiments were conducted to test the robot's docking ability and the accuracy of the proposed EKF. In order to quantify the accuracy of the EKF, the true positions of each module and connector are measured using an ultrasonic position sensor (CMS10, Zebris) whose positional error is typically less than 2 mm. Two ultrasonic emitters are attached to the body of the module to find the orientation of the module and one ultrasonic emitter is attached to the connector.

Three different docking experiments were conducted. In the first case, two modules are initially facing each other. For the second case, the modules are initially misaligned by 90°. Finally, in the third case, orientation error is introduced by manually rotating one module before the EKF is applied.

Modules Facing Each Other

The two modules are initially orientated so that they are facing each other, and then the docking process is performed. Figure 13 shows the experiment progression. Red circles represent the ultrasonic sensor attached to the connector and each blue box represents the module obtained from two ultrasonic sensors. Initially, the two modules are facing each other as shown in Figure 13a. When docking is initiated, Module 1 rotates to align with Module 2 (Figure 13b). After Module 1 is aligned, Module 2 is aligned (Figure 13c). Once both modules are aligned, Module 1 moves forward until the distance estimation from EKF becomes 120 mm (Figure 13d), which is the length of the two connectors. While Module 1 moves forward, the EKF is activated to estimate the distance between the two modules. When the estimated distance is less than 120 mm, Module 1 moves backwards to check if the modules are docked. If docking is successful, Module 1 will drag Module 2, and the distance between the two modules remains 120 mm. In this case, IR receiver will measure high signal strength. Otherwise, only Module 1 moves backward, and the IR signal strength measurements will be low. Figure 13e shows that Module 1 moves backwards with Module 2 because docking was successful. Therefore, the IR signal strength is high, and the system concludes that the two modules are successfully docked, and the modules stop (Figure 13f). A video of experiments can be found in the supplementary file.

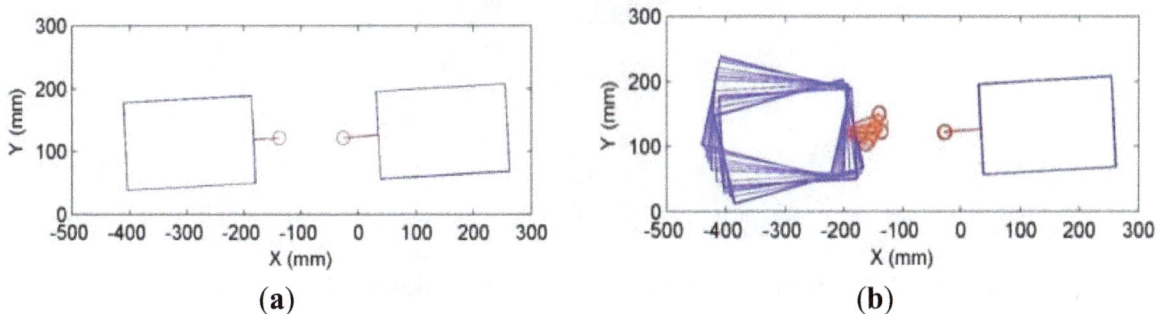

(a)   (b)

Figure 13. *Cont.*

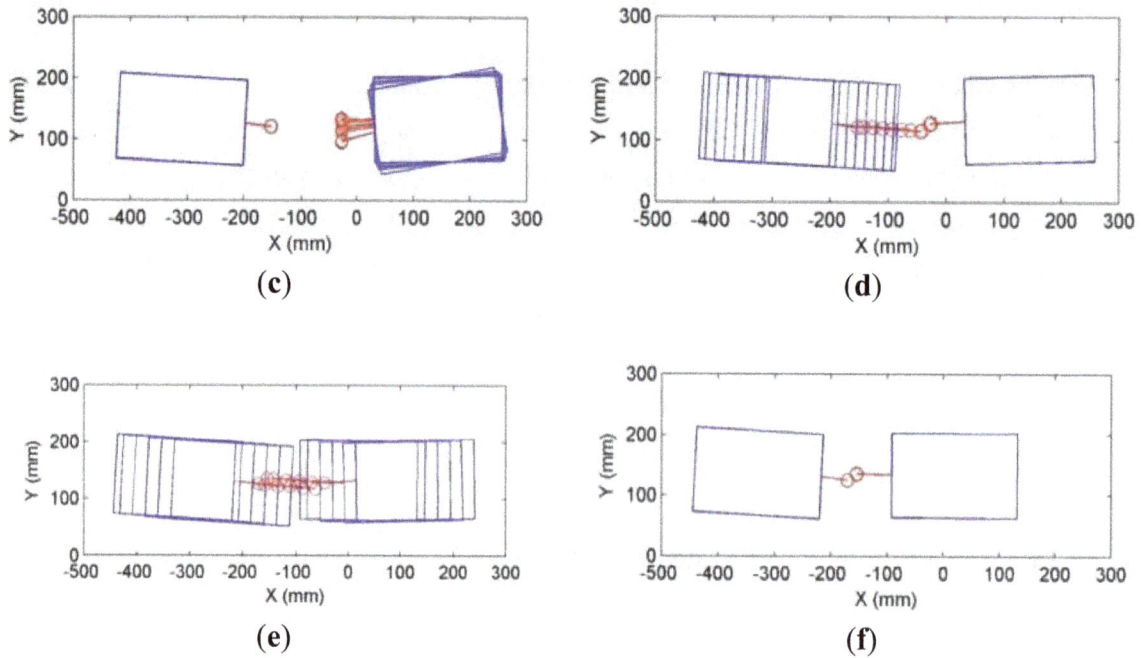

Figure 13. Experimental results when two modules are initially facing each other. (**a**) initial state, (**b**) Alignment of Module 1, (**c**) Alignment of Module 2, (**d**) Module 1 moves forwards for docking, (**e**) Module 1 moves backwards for checking successful docking, and (**f**) Docking process completed.

The EKF is activated when Module 1 starts moving towards Module 2 (Figure 13d) and is deactivated when the distance, L, is less than 120 mm. The distance estimation error of EKF is shown in Figure 14. The error graph shows that the distance estimation error is less than 4 mm while Module 1 is moving forward for docking. The magnitude of the distance error is within the tolerance of the magnetic connector.

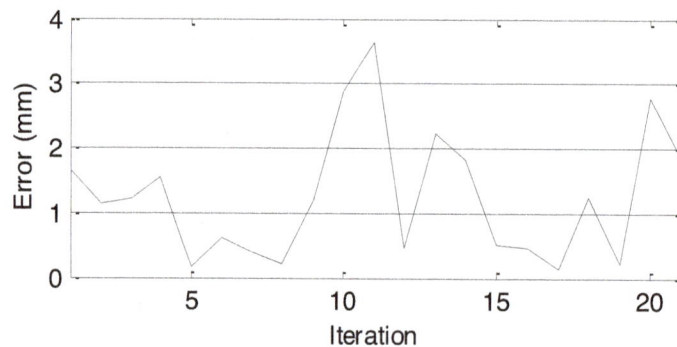

Figure 14. EKF distance error of when two modules are facing each other.

4.3. Docking Modules at 90° Angle

Initially, Module 1 faces Module 2 with an angle of 90° as shown in Figure 15a. First, Module 1 keeps rotating until it roughly aligns with Module 2. Then, Module 1 rotates for the fine-alignments (Figure 15b). Next, Module 2 is aligned. After the alignments, Module 1 moves forward for docking (Figure 15c). However, Figure 15c shows offset between two modules, which means there is some error in emitter angle (θe) and receiver angle (θr). This also results in distance error. Therefore, even when Module 1 docks with Module 2 successfully as shown in Figure 15c, it keeps moving forward and pushes the connector of Module 2. Thus, the connector of Module 2 is tilted a bit further in Figure 15d compared to Figure 15c. When distance estimation is 120 mm, Module 1 moves back to check if docking is successful (Figure 15e). Since docking is completed successfully, Module 1 stops (Figure 15f). This experiment shows that magnets pull each other and successfully dock even with a small offset error because the universal joint and suction cup (Figure 2) make the magnets move freely. The error analysis in Figure 16 shows that the error is similar as the previous case.

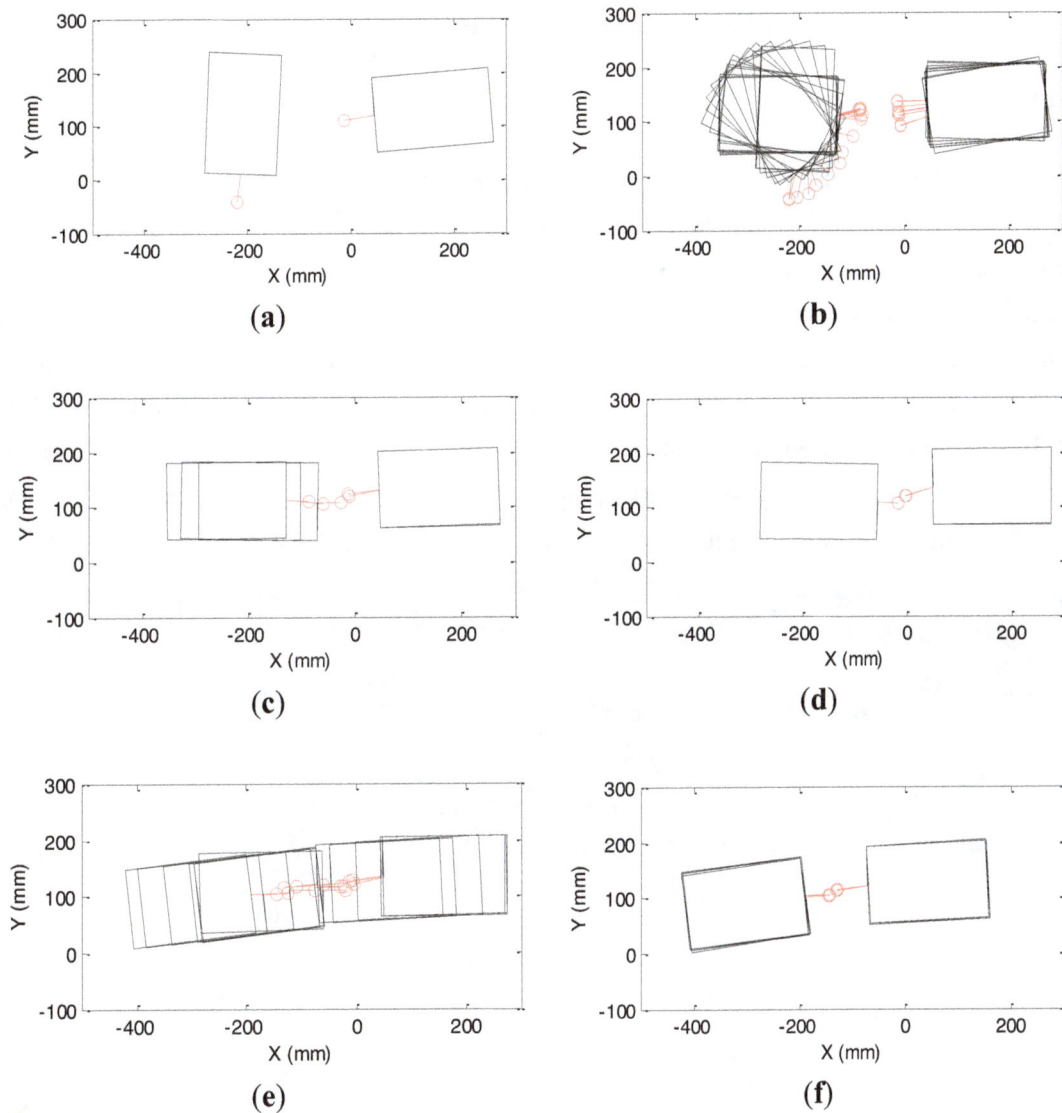

Figure 15. Experimental results when two modules are misaligned roughly 90°. (**a**) Initial state, (**b**) Alignment of Module 1and Module 2, (**c**) Module 1 moves forwards for docking, (**d**) Tow modules dock, (**e**) Module 1 moves backwards for checking successful docking, and (**f**) Docking process completed.

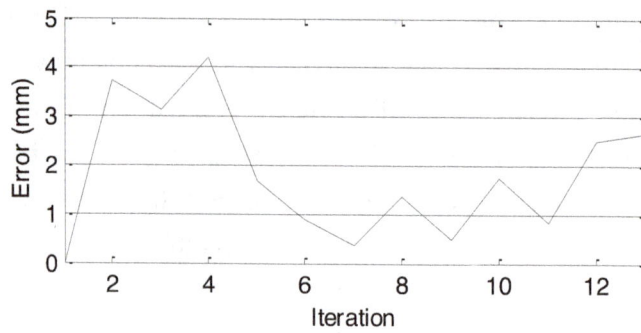

Figure 16. EKF distance error when misalignment is 90°.

Incorrect Initial State Estimation

The performance of the docking algorithm is evaluated in the presence of the initial state estimation errors. For this experiment, Module 1 is lifted up and manually rotated before EKF is applied. Therefore, errors are introduced to the initial state estimations. Figure 17 shows the progress of this experiment. The initial states of the two modules are shown in Figure 17a. First, Module 1 is aligned (Figure 17b) and then Module 2 is rotated to be aligned (Figure 17c). Before Module 2 is completely aligned, Module 1 is manually rotated to introduce errors. Figure 17d shows that Module 1 was initially aligned with Module 2, but manually rotated. Then, Module 1 moves forward as shown in Figure 17e. The distance estimation error during this step is shown in Figure 18a. The figure shows that the initial distance error is greater than 40 mm due to incorrect initial estimation.

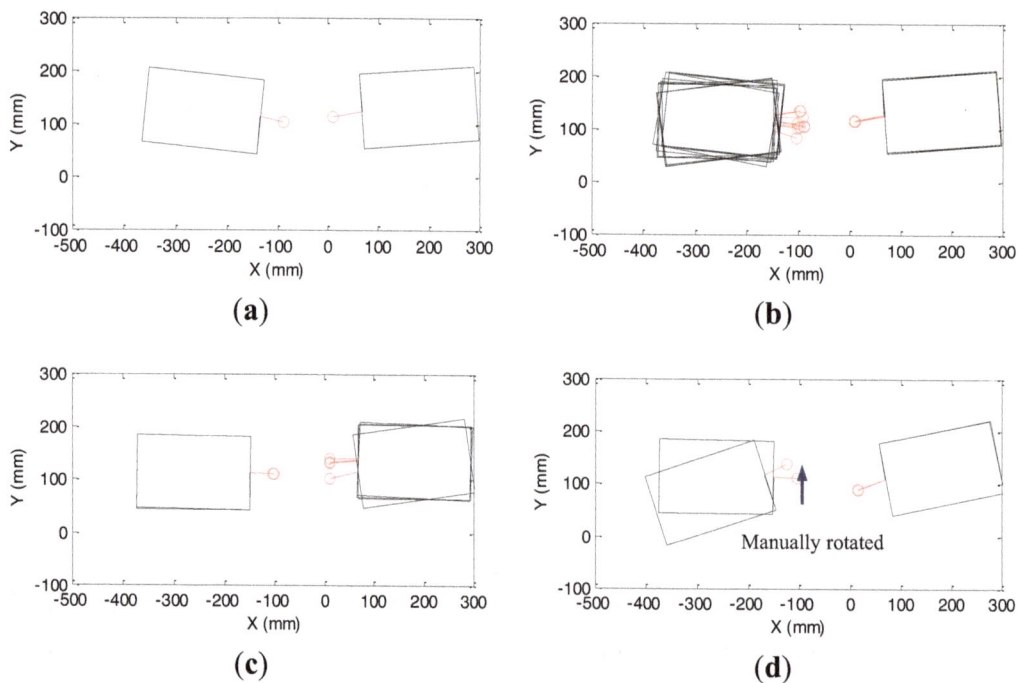

(a)

(b)

(c)

(d)

Figure 17. *Cont.*

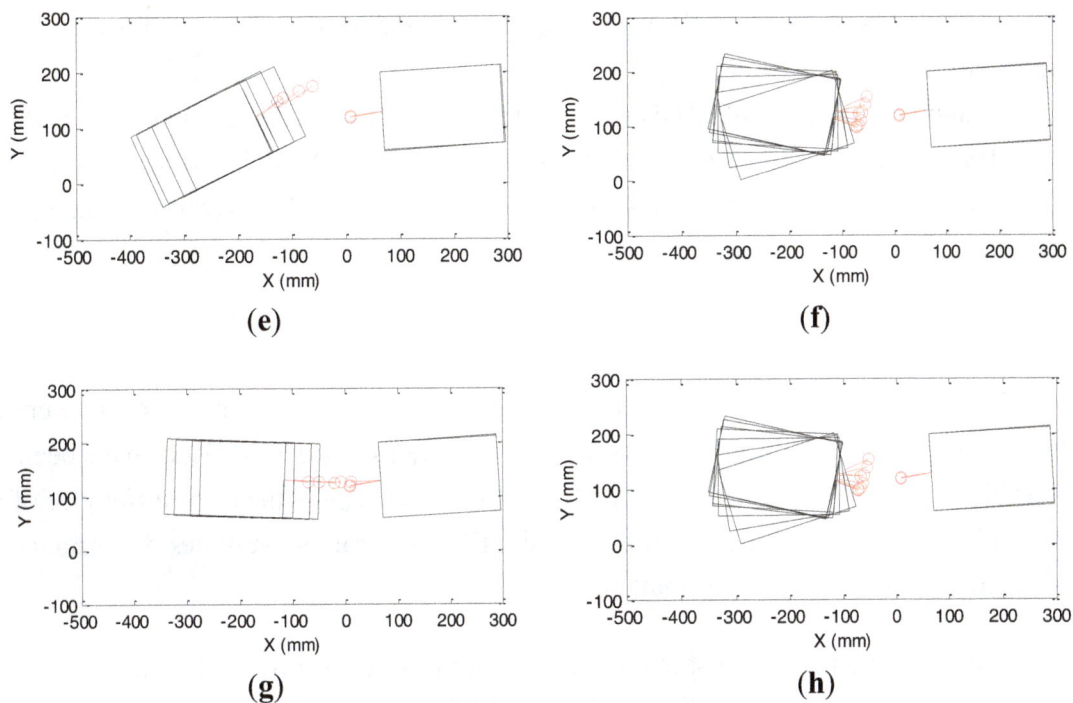

(e)   (f)

(g)   (h)

Figure 17. Experimental results: Module 1 is manually rotated just before KF is initialized so that initial estimations of distance and emitter and receiver angles are incorrect. (**a**) Initial state, (**b**) Alignment of Module 1, (**c**) Alignment of Module 2, (**d**) Module 1 was rotated manually to introduce orientation error, (**e**) Module 1 moves forwards for docking, (**f**) Module 1 rotates again for realignment, (**g**) Module 1 moves forwards for docking, and (**h**) Module 1 moves backwards for checking successful docking.

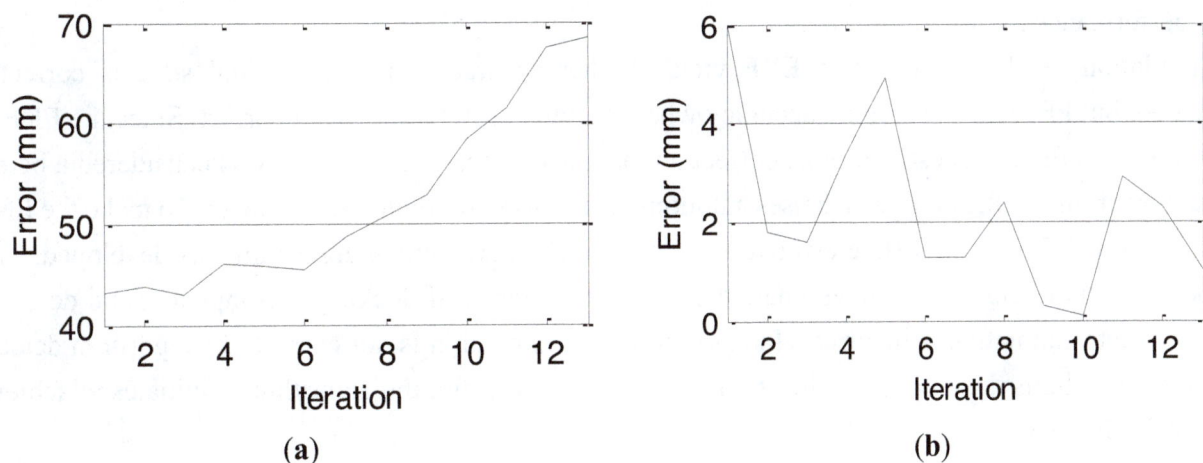

(a)   (b)

Figure 18. EKF distance error with incorrect initial state estimation (**a**) before re-initialization (**b**) after re-initialization.

Also, as Module 1 moves forwards, the distance error keeps increasing. Figure 17f shows that Module 1 stops and rotates for realignment because the algorithm identified that initial estimations were not correct. After Module 1 is realigned, EKF is initialized again and Module 1 moves forward for docking, as shown in Figure 17g.

Figure 17g is shown in Figure 18b. The error graph shows that the distance error after realignment is reduced to the level of previous two scenarios. Figure 17h shows that Module 1 drags Module 2 to check for successful docking. This experiment shows that the proposed algorithm is robust enough to dock two modules even if high initial estimation error exists.

4.4. Repeatability

To prove the repeatability of the experimental results, the above three experiments were repeated five times each and the average distance errors at docking were found. These results have been tabulated in Table 5. Table 5 shows that the errors are small in all three cases, and the error difference of the three cases is insignificant. The results demonstrate that the EKF accurately estimates the distance between the two robots in all three cases with a small error.

Table 5. Average state estimation distance error of three different scenarios.

	Facing Each Other	90° Misalignment	With Error
Distance error (mm)	2.11	1.53	1.89

5. Conclusions

This paper presents a novel docking system for modular mobile self-reconfigurable robots using low cost sensors. Low cost IR sensors and encoders are utilized to estimate the distance and the orientation of the modules. The signal strength of the IR sensor was modeled using emitter and receiver angles as well as the distance. Based on the IR sensor and motion models, an EKF and a PF were developed to estimate the distance and orientation more accurately. Both state estimators were simulated to compare their performance in this application.

Simulation results showed that EKF yields higher accuracy when the initial state is correctly estimated, but PF results in a better accuracy when the initial estimation is not correct. Since the EKF is accurate when the initial estimation is correct and is much faster than PF, EKF was considered a better estimator for this application, which uses autonomous robots with limited CPU power. To make the EKF work even with high initial state estimation error, an efficient docking algorithm was developed. The proposed docking algorithm can estimate the initial state, detect if docking is completed, and detect if the EKF state estimation is incorrect. In case the initial estimation is not correct, the algorithm detects the error and aligns the module again. If docking is not successful, the algorithm re-initiates to achieve successful docking.

The proposed system with the EKF was experimentally tested in three different case studies. When the initial estimation is correctly identified, the EKF can correctly estimate the position, which allows successful docking. When the initial state is not correctly estimated, the distance estimation diverges from the true values. However, the proposed algorithm can identify the error, and make the modules re-align. After the re-alignment, the distance estimation of EKF is accurate and docking can be successfully completed. These experiments demonstrate the robustness of the proposed docking system.

The experimental results demonstrate that the proposed system can accurately estimate the distance between the two modules, and make the robots dock successfully even when the initial distance between robots is not calculated correctly. Therefore, our contribution of this work lies in the development of a robust, cost-effective, and reliable docking mechanism with accurate distance estimations.

IR sensors present a challenge for long range docking system because the signal strength decreases quadratically over distance. IR sensors can be used for a long range, but they require high power consumption. Thus, for long-range applications, another sensing system such as RF can be exploited to steer the modules closer to each other and then IR sensing system can be used for docking when the modules are in close proximity.

Acknowledgments

Financial support from the Natural Sciences and Engineering Research Council of Canada is appreciated.

Author Contributions

All authors contributed to this work.

Conflicts of Interest

The authors declare no conflict of interest.

References

1. Yim, M.; Zhang, Y.; Duff, D. Modular robots. *IEEE Spectrum* **2002**, *39*, 30–34.
2. Moubarak, P.; Ben-Tzvi, P. Modular and Reconfigurable Mobile Robotics. *J. Robot. Auton. Syst.* **2012**, *60*, 1648–1663.
3. Yim, M.; Shen, W.; Salemi, B.; Rus, D.; Moll, M.; Lipson, H.; Klavins, E.; Chirikjian, G.S. Modular self-reconfigurable robot systems. *IEEE Robot. Autom. Mag.* **2007**, *14*, 43–52.
4. Thomas, T.; Wagner, F.; Witkowski, F. Modular mobile robot platform for research and academic applications in embedded systems. In *Advances in Autonomous Robotics*; Springer: Berlin/Heidelberg, Germany, 2012; Volume 7429, pp. 270–278.
5. Shoval, S.; Borenstein, J. Measuring the Relative Position and Orientation between Two Mobile Robots with Binaural Sonar. In Proceedings of the American Nuclear Society 9th International Topical Meeting on Robotics and Remote Systems, Seattle, DC, USA, 4–8 March 2001.
6. Wang, W.; Li, Z.; Yu, W.; Zhang, J. An autonomous docking method based on ultrasonic sensors for self-reconfigurable mobile robot. In Proceedings of the IEEE International Conference on Robotics and Biomimetics, Guilin, China, 19–23 December 2009.
7. Kim, M.; Chong, N.Y.; Yu, W. Robust DOA estimation and target docking for mobile robots. *Intell. Serv. Robot.* **2009**, *2*, 41–51.
8. Kim, Y.-H.; Lee, S.-W.; Yang, H.S.; Shell, D.A. Toward autonomous robotic containment booms: Visual servoing for robust inter-vehicle docking of surface vehicles. *Intell. Serv. Robot.* **2012**, *5*, 1–18.

9. Murata, S.; Kakomura, K.; Kurokawa, H. Docking Experiments of a Modular Robot by Visual Feedback. In Proceedings of the EEE/RSJ International Conference on Intelligent Robots and Systems, Beijing, China, 9–15 October 2006.

10. Fidan, B.; Gazi, V.; Zhai, S.; Cen, N.; Karatas, E. Single-View Distance-Estimation-Based Formation Control of Robotic Swarms. *IEEE Trans. Ind. Electron.* **2013**, *60*, 5781–5791.

11. Liu, W.; Winfield, A. Implementation of an IR approach for autonomous docking in a self-configurable robotics system. In Proceedings of the Towards Autonomous Robotic Systems, Londonderry, UK, 31 August–2 September 2009.

12. Zhang, Y.; Roufas, K.; Eldershaw, C.; Yim, M.; Duff, D. Sensor Computations in Modular Self Reconfigurable Robots. In Proceedings of the International Symposium on Experimental Robotics 2002, Sant'Angelo d'Ischia, Italy, 8–11 July 2002.

13. Rubenstein, M.; Payne, K.; Will, P.; Shen, W.-M. Docking among independent and autonomous CONRO self-reconfigurable robots. In Proceedings of the 2004 IEEE International Conference on Robotics and Automation, Marina del Rey, CA, USA, 26 April–1 May 2004.

14. Mao, L.; Chen, J.; Li, Z.; Zhang, D. Relative Localization Method of Multiple Micro Robots Based on Simple Sensors. *Int. J. Adv. Robot. Syst.* **2013**, *10*, 1–9.

15. Benet, G.; Blanes, F.; Simó, J.E.; Pérez, P. Using infrared sensors for distance measurement in mobile robots. *Robot. Auton. Syst.* **2002**, *40*, 255–266.

16. Roufas, K.; Zhang, Y.; Duff, D.; Yim, M. Six Degree of Freedom Sensing for Docking Using IR LED Emitters and Receivers. In *Experimental Robotics* VII, Springer: Berlin/Heidelberg, Germany, 2001; pp. 91–100.

17. Kiriy, E.; Buehler, M. *Three-State Extended Kalman Filter for Mobile Robot Localization*; Technical Report; McGill University: Montreal, QC, Canada, 2002.

18. Kim, S.J.; Kim, B.K. Dynamic Ultrasonic Hybrid Localization System for Indoor Mobile Robots. *IEEE Trans. Ind. Electron.* **2013**, *60*, 4562–4573.

19. Wang, C.M. Localization estimation and uncertainty analysis for mobile robots. In Proceedings of the IEEE International Conference on Robotics and Automation, Philadelphia, PA, USA, 24–29 April 1988.

20. Chang, D.-C.; Fang, M.-W. Bearing-Only Maneuvering Mobile Tracking With Nonlinear Filtering Algorithms in Wireless Sensor Networks. *IEEE Syst. J.* **2014**, *8*, 160–170.

21. Conti, A.; Dardari, D.; Guerra, M.; Mucchi, L.; Win, M. Experimental Characterization of Diversity Navigation. *IEEE Syst. J.* **2014**, *8*, 115–124.

8

The Role of Visibility in Pursuit/Evasion Games

Athanasios Kehagias [1,*]**, Dieter Mitsche** [2] **and Paweł Prałat** [3]

[1] Department of Electrical and Computer Engineering, Aristotle University, GR 54248, Thessaloniki, Greece
[2] Laboratoire J. A. Dieudonné, UMR CNRS-UNS No 7351, Université de Nice Sophia-Antipolis, Parc Valrose 06108 Nice Cedex 2, France; E-Mail: dmitsche@unice.fr
[3] Department of Mathematics, Ryerson University, 350 Victoria St., Toronto, ON, M5B 2K3, Canada; E-Mail: pralat@ryerson.ca

* Author to whom correspondence should be addressed; E-Mail: kehagiat@gmail.com

External Editor: Wenjie Dong

Abstract: The *cops-and-robber* (CR) game has been used in mobile robotics as a discretized model (played on a graph G) of *pursuit/evasion* problems. The "classic" CR version is a *perfect information game*: the cops' (pursuer's) location is always known to the robber (evader) and vice versa. Many variants of the classic game can be defined: the robber can be *invisible* and also the robber can be either *adversarial* (tries to avoid capture) or *drunk* (performs a random walk). Furthermore, the cops and robber can reside in either nodes or edges of G. Several of these variants are relevant as models or robotic pursuit/evasion. In this paper, we first define carefully several of the variants mentioned above and related quantities such as the *cop number* and the *capture time*. Then we introduce and study the *cost of visibility* (COV), a quantitative measure of the increase in difficulty (from the cops' point of view) when the robber is invisible. In addition to our theoretical results, we present algorithms which can be used to compute capture times and COV of graphs which are analytically intractable. Finally, we present the results of applying these algorithms to the numerical computation of COV.

Keywords: mobile robotics; robot coordination; pursuit/evasion

1. Introduction

Pursuit/evasion (PE) and related problems (search, tracking, surveillance) have been the subject of extensive research in the last fifty years and much of this research is connected to mobile robotics [1]. When the environment is represented by a graph (for instance, a floorplan can be modeled as a graph, with nodes corresponding to rooms and edges corresponding to doors; similarly, a maze can be represented by a graph with edges corresponding to tunnels and nodes corresponding to intersections), the original PE problem is reduced to a *graph game* played between the pursuers and the evader.

In the current paper, inspired by Isler and Karnad's recent work [2], we study the role of *information* in *cops-and-robber* (CR) games, an important version of graph-based PE. By "information" we mean specifically the players' *location*. For example, we expect that when the cops know the robber's location they can do better than when the robber is "invisible". Our goal is to make precise the term "better".

Reviews of the graph theoretic CR literature appear in [3–5]. In the "classical" CR variant [6] it is assumed that the cops always know the robber's location and vice versa. The "invisible" variant, in which the cops cannot see the robber (but the robber always sees the cops) has received less attention in the graph theoretic literature; among the few papers which treat this case we mention [2,7–9] and also [10] in which *both* cops and robber are invisible.

Both the visible and invisible CR variants are natural models for discretized robotic PE problems; the connection has been noted and exploited relatively recently [2,8,11]. If it is further assumed that the robber is not actively trying to avoid capture (the case of *drunk* robber) we obtain a *one-player* graph game; this model has been used quite often in mobile robotics [12–16] and especially (when assuming random robber movement) in publications such as [17–21], which utilize *partially observable Markov decision processes* (POMDP, [22–24]). For a more general overview of pursuit/evasion and search problems in robotics, the reader is referred to [1]; some of the works cited in this paper provide a useful background to the current paper. Finally, several related works have also been published in the Distributed Algorithms community [25–27].

This paper is structured as follows. In Section 2 we present preliminary material, notation and the definition of the "classical" CR game; we also introduce several node and edge CR *variants*. In Section 3 we define rigorously the *cop number* and *capture time* for the classical CR game and the previously introduced CR variants. In Section 4 we study the *cost of visibility* (COV). In Section 5 we present algorithms which compute capture time and optimal strategies for several CR variants. In Section 6 we further study COV using computational experiments. Finally, in Section 7 we summarize and present our conclusions.

2. Preliminaries

2.1. Notation

1. We use the following notations for sets: $\mathbb{N}$ denotes $\{1, 2, \ldots\}$; $\mathbb{N}_0$ denotes $\{0, 1, 2, \ldots\}$; $[K]$ denotes $\{1, \ldots, K\}$; $A - B = \{x : x \in A, x \notin B\}$; $|A|$ denotes the *cardinality* of A (*i.e.*, the number of its elements).

2. A *graph* $G = (V, E)$ consists of a *node set* V and an *edge set* E, where every $e \in E$ has the form $e = \{x, y\} \subseteq V$. In other words, we are concerned with finite, undirected, simple graphs; in addition we will always assume that G is connected and that G contains n nodes: $|V| = n$. Furthermore, we will assume, without loss of generality, that the node set is $V = \{1, 2, ..., n\}$. We let $V^K = \underbrace{V \times V \times \ldots \times V}_{K \text{ times}}$. We also define $V_D^2 \subseteq V^2$ by $V_D^2 = \{(x, x) : x \in V\}$ (it is the set of "diagonal" node pairs).

3. A *directed graph (digraph)* $G = (V, E)$ consists of a *node set* V and an *edge set* E, where every $e \in E$ has the form $e = (x, y) \in V \times V$. In other words, the edges of a digraph are *ordered* pairs.

4. In graphs, the *(open) neighborhood* of some $x \in V$ is $N(x) = \{y : \{x, y\} \in E\}$; in digraphs it is $N(x) = \{y : (x, y) \in E\}$. In both cases, the *closed neighborhood* of x is $N[x] = N(x) \cup \{x\}$.

5. Given a graph $G = (V, E)$, its *line graph* $L(G) = (V', E')$ is defined as follows: the node set is $V' = E$, i.e., it has one node for every edge of G; the edge set is defined by having the nodes $\{u, v\}, \{x, y\} \in V'$ connected by an edge $\{\{u, v\}, \{x, y\}\}$ if and only if $|\{u, v\} \cap \{x, y\}| = 1$ (i.e., if the original edges of G are adjacent).

6. We will write $f(n) = o(g(n))$ if and only if $\lim_{n \to \infty} \frac{f(n)}{g(n)} = 0$. Note that in this *asymptotic* notation n denotes the parameter with respect to which asymptotics are considered. So in later sections we will write $o(n), o(M)$ etc.

2.2. The CR Game Family

The "classical" CR game can be described as follows. Player C controls K cops (with $K \geq 1$) and player R controls a single robber. Cops and robber are moved along the edges of a graph $G = (V, E)$ in discrete time steps $t \in \mathbb{N}_0$. At time t, the robber's location is $Y_t \in V$ and the cops' locations are $X_t = (X_t^1, X_t^2, \ldots, X_t^K) \in V^K$ (for $t \in \mathbb{N}_0$ and $k \in [K]$). The game is played in *turns*; in the 0-th turn first C places the cops on nodes of the graph and then R places the robber; in the t-th turn, for $t > 0$, *first* C moves the cops to X_t and *then* R moves the robber to Y_t. Two types of moves are allowed: (a) sliding along a single edge and (b) staying in place; in other words, for all t and k, either $\{X_{t-1}^k, X_t^k\} \in E$ or $X_{t-1}^k = X_t^k$; similarly, $\{Y_{t-1}, Y_t\} \in E$ or $Y_{t-1} = Y_t$. The cops win if they *capture* the robber, i.e., if there exist $t \in \mathbb{N}_0$ and $k \in [K]$ such that $Y_t = X_t^k$; the robber wins if for all $t \in \mathbb{N}_0$ and $k \in [K]$ we have $Y_t \neq X_t^k$. In what follows we will describe these eventualities by the following "shorthand notation": $Y_t \in X_t$ and $Y_t \notin X_t$ (i.e., in this notation we consider X_t as a *set* of cop positions).

In the classical game both C and R are *adversarial*: C plays to effect capture and R plays to avoid it. But there also exist "drunk robber" versions, in which the robber simply performs a *random walk* on G such that, for all $\forall u, v \in V$ we have

$$\Pr(Y_0 = u) = \frac{1}{n} \quad \text{and} \quad \Pr(Y_{t+1} = u | Y_t = v) = \begin{cases} \frac{1}{|N(v)|} & \text{if and only if } u \in N(v) \\ 0 & \text{otherwise} \end{cases} \tag{1}$$

In this case we can say that no R player is present (or, following a common formulation, we can say that the R player is "Nature").

If an R player exists, the cops' locations are always known to him; on the other hand, the robber can be either *visible* (his location is known to C) or *invisible* (his location is unknown). Hence we have four different CR variants, as detailed in the following Table 1.

Table 1. Four variants of the CR game.

Adversarial Visible Robber	av-CR
Adversarial Invisible Robber	ai-CR
Drunk Visible Robber	dv-CR
Drunk Invisible Robber	di-CR

In all of the above CR variants both cops and robber move from node to node. This is a good model for entities (e.g., robots) which move from room to room in an indoor environment. There also exist cases (for example moving in a maze or a road network) where it makes more sense to assume that both cops and robber move from edge to edge. We will call the classical version of the edge CR game *edge av-CR*; it has attracted attention only recently [28]. Edge ai-CR, dv-CR and di-CR variants are also possible, in analogy to the node versions listed in the Table. Each of these cases can be reduced to the corresponding node variant, with the edge game taking place on the line graph $L(G)$ of G.

3. Cop Number and Capture Time

Two graph parameters which can be obtained from the av-CR game are the *cop number* and the *capture time*. In this section we will define these quantities in game theoretic terms (while this approach is not common in the CR literature, we believe it offers certain advantages in clarity of presentation) and also consider their extensions to other CR variants. Before examining each of these CR variants in detail, let us mention a particular modification which we will apply to all of them. Namely, we assume that (every variant of) the CR game is played for an *infinite number of rounds*. This is obviously the case if the robber is never captured; but we also assume that, in case the robber is captured at some time t^*, the game continues for $t \in \{t^* + 1, t^* + 2, \ldots\}$ with the following restriction: for all $t \geq t^*$, we have $Y_t = X_t^{k^*}$ (where k^* is the number of cop who effected the capture). This modification facilitates the game theoretic analysis presented in the sequel; intuitively, it implies that after capture, the k^*-th cop forces the robber to "follow" him.

3.1. The Node av-CR Game

We will define cop number and capture time in game theoretic terms. To this end we must first define *histories* and *strategies*.

A particular instance of the CR game can be fully described by the sequence of cops and robber locations; these locations are fully determined by the C and R moves. So, if we let $x_t \in V^K$ (resp. $y_t \in V$) denote the nodes into which C (resp. R) places the cops (resp. the robber) at time t, then a *history* is a sequence $x_0 y_0 x_1 y_1 \ldots$. Such a sequence can have finite or infinite length; we denote the set of all finite length histories by $H_*^{(K)}$; note that there exists an infinite number of finite length sequences. By convention $H_*^{(K)}$ also includes the *zero-length* or *null* history, which is the *empty sequence* (this corresponds to the beginning of the game, when neither player has made a move, just before C places the cops on G), denoted by λ. Finally, we denote the set of all infinite length histories by $H_\infty^{(K)}$.

Since both cops and robber are visible and the players move sequentially, av-CR is a game of *perfect information*; in such a game C loses nothing by limiting himself to *pure* (*i.e.*, deterministic) *strategies* [29]. A pure cop strategy is a function $s_C : H_*^{(K)} \to V^K$; a pure robber strategy is a function $s_R : H_*^{(K)} \to V$. In both cases the idea is that, given a finite length history, the strategy produces the next cop or robber move (note the dependence on K, the number of cops); for example, when the robber strategy s_R receives the input x_0, it will produce the output $y_0 = s_R(x_0)$; when it receives $x_0 y_0 x_1$, it will produce $y_1 = s_R(x_0 y_0 x_1)$ and so on. We will denote the set of all legal cop strategies by $\mathbf{S}_C^{(K)}$ and the set of all legal robber strategies by $\mathbf{S}_R^{(K)}$; a strategy is "legal" if it only provides moves which respect the CR game rules. The set $\widetilde{\mathbf{S}}_C^{(K)} \subseteq \mathbf{S}_C^{(K)}$ (resp. $\widetilde{\mathbf{S}}_R^{(K)} \subseteq \mathbf{S}_R^{(K)}$) is the set of *memoryless* legal cop (resp. robber) strategies, *i.e.*, strategies which only depend only on the current cops and robber positions; we will denote the memoryless strategies by Greek letters, e.g., σ_C, σ_R *etc.* In other words

$$\sigma_C \in \widetilde{\mathbf{S}}_C^{(K)} \Rightarrow [\forall t : x_{t+1} = \sigma_C(x_0 y_0 \ldots . x_t y_t) = \sigma_C(x_t y_t)]$$

$$\sigma_R \in \widetilde{\mathbf{S}}_R^{(K)} \Rightarrow [\forall t : y_{t+1} = \sigma_R(x_0 y_0 \ldots . x_t y_t x_{t+1}) = \sigma_R(y_t x_{t+1})]$$

It seems intuitively obvious that both C and R lose nothing by playing with memoryless strategies (*i.e.*, computing their next moves based on the current position of the game, not on its entire history). This is true but requires a proof. One approach to this proof is furnished in [30,31]. But we will present another proof by recognizing that the CR game belongs to the extensively researched family of *reachability games* [32,33].

A reachability game is played by two players (Player 0 and Player 1) on a *digraph* $\overline{G} = (\overline{V}, \overline{E})$; each node $v \in \overline{V}$ is a *position* and each edge is a *move*; *i.e.*, the game moves from node to node (position) along the edges of the digraph. The game is described by the tuple $(\overline{V}_0, \overline{V}_1, \overline{E}, \overline{F})$, where $\overline{V}_0 \cup \overline{V}_1 = \overline{V}$, $\overline{V}_0 \cap \overline{V}_1 = \emptyset$ and $\overline{F} \subseteq \overline{V}$. For $i \in \{0, 1\}$, $\overline{V}_i$ is the set of positions (nodes) from which the i-th Player makes the next move; the game terminates with a win for Player 0 if and only if a move takes place into a node $v \in \overline{F}$ (the *target set* of Player 0); if this never happens, Player 1 wins. Here is a more intuitive description of the game: each move consists in sliding a *token* from one digraph node to another, along an edge; the i-th player slides the token if and only if it is currently located on a node $v \in \overline{V}_i$ ($i \in \{0, 1\}$); Player 0 wins if and only if the token goes into a node $u \in \overline{F}$; otherwise Player 1 wins. The following is well known [32,33].

Theorem 1. *Let* $(\overline{V}_0, \overline{V}_1, \overline{E}, \overline{F})$ *be a reachability game on the digraph* $\overline{D} = (\overline{V}, \overline{E})$. *Then* $\overline{V}$ *can be partitioned into two sets* $\overline{W}_0$ *and* $\overline{W}_1$ *such that (for* $i \in \{0, 1\}$*) player* i *has a* memoryless *strategy* σ_i *which is winning whenever the game starts in* $u \in \overline{W}_i$.

We can convert the av-CR game with K cops to an equivalent reachability game which is played on the *CR game digraph*. In this digraph every node corresponds to a *position* of the original CR game; a (directed) edge from node u to node v indicates that it is possible to get from position u to position v in a single move. The CR game digraph has three types of nodes.

1. Nodes of the form $u = (x, y, p)$ correspond to positions (in the original CR game) with the cops located at $x \in V^K$, the robber at $y \in V$ and player $p \in \{C, R\}$ being next to move.

2. There is single node $u = (\lambda, \lambda, C)$ which corresponds to the starting position of the game: neither the cops nor the robber have been placed on G; it is C's turn to move (recall that λ denotes the empty sequence).

3. Finally, there exist n nodes of the form $u = (x, \lambda, R)$: the cops have just been placed in the graph (at positions $x \in V^K$) but the robber has not been placed yet; it is R's turn to move.

Let us now define

$$\overline{V}_0^{(K)} = \left\{ (x, y, C) : x \in V^K \cup \{\lambda\}, y \in V \cup \{\lambda\} \right\}$$
$$\overline{V}_1^{(K)} = \left\{ (x, y, R) : x \in V^K \cup \{\lambda\}, y \in V \cup \{\lambda\} \right\}$$
$$\overline{V}^{(K)} = \overline{V}_0^{(K)} \cup \overline{V}_1^{(K)}$$

and let $\overline{E}^{(K)}$ consist of all pairs (u, v) where $u, v \in \overline{V}^{(K)}$ and the move from u to v is legal. Finally, we recognize that C's target set is

$$\overline{F}^{(K)} = \left\{ (x, y, p) : x \in V^K, y \in (V \cap x), p \in \{C, R\} \right\}$$

i.e., the set of all positions in which the robber is in the same node as at least one cop.

With the above definitions, we have mapped the classical CR game (played with K cops on the graph G) to the reachability game $\left(\overline{V}_0^{(K)}, \overline{V}_1^{(K)}, \overline{E}^{(K)}, \overline{F}^{(K)} \right)$. By Theorem 1, Player i (with $i \in \{0, 1\}$) will have a *winning set* $\overline{W}_i^{(K)} \subseteq \overline{V}^{(K)}$, *i.e.*, a set with the following property: whenever the reachability game starts at some $u \in \overline{W}_i^{(K)}$, then Player i has a winning strategy (it may be the case, for specific G and K that either of $\overline{W}_0^{(K)}, \overline{W}_1^{(K)}$ is empty). Recall that in our formulation of CR as a reachability game, Player 0 is C. In reachability terms, the statement "C has a winning strategy in the classical CR game" translates to "$(\lambda, \lambda, C) \in \overline{W}_0^{(K)}$" and, for a given graph G, the validity of this statement will in general depend on K. It is clear that $\overline{W}_0^{(K)}$ is increasing with K:

$$K_1 \leq K_2 \Rightarrow \overline{W}_0^{(K_1)} \subseteq \overline{W}_0^{(K_2)} \tag{2}$$

It is also also clear that

$$\text{``}(\lambda, \lambda, C) \in \overline{W}_0^{(|V|)}\text{''} \text{ is true for every } G = (V, E) \tag{3}$$

because, if C has $|V|$ cops, he can place one in every $u \in V$ and win immediately. In fact, for $K = |V|$, we have $\overline{W}_0^{(|V|)} = \overline{V}^{(K)}$, because from every position (x, y, p), C can move the cops so that one cop resides in each $u \in V$, which guarantees immediate capture.

Based on Equations (2) and (3) we can define the *cop number* of G to be the minimum number of cops that guarantee capture; more precisely we have the following definition (which is equivalent to the "classical" definition of cop number [34]).

Definition 1. *The* cop number *of G is*

$$c(G) = \min \left\{ K : (\lambda, \lambda, C) \in \overline{W}_0^{(K)} \right\}$$

While a cop winning strategy s_C guarantees that the token will go into (and remain in) $\overline{F}^{(K)}$, we still do not know how long it will take for this to happen. However, it is easy to prove that, if $K \geq c(G)$ and C uses a *memoryless* winning strategy, then no game position will be repeated until capture takes place. Hence the following holds.

Theorem 2. *For every G, let $K \geq c(G)$ and consider the CR game played on G with K cops. There exists a a memoryless cop winning strategy σ_C and a number $\overline{T}(K; G) < \infty$ such that, for every robber strategy s_R, C wins in no more than $\overline{T}(K; G)$ rounds.*

Let us now turn from winning to *time optimal* strategies. To define these, we first define the *capture time*, which will serve as the CR *payoff function*.

Definition 2. *Given a graph G, some $K \in \mathbb{N}$ and strategies $s_C \in \mathbf{S}_C^{(K)}$, $s_R \in \mathbf{S}_R^{(K)}$ the av-CR capture time is defined by*

$$T^{(K)}(s_C, s_R|G) = \min\left\{t : \exists k \in [K] \text{ such that } Y_t = X_t^k\right\} \tag{4}$$

in case capture never takes place, we let $T^{(K)}(s_C, s_R|G) = \infty$.

We will assume that R's *payoff* is $T^{(K)}(s_C, s_R|G)$ and C's payoff is $-T^{(K)}(s_C, s_R|G)$ (hence av-CR is a *two-person zero-sum game*). Note that capture time (i) obviously depends on K and (ii) for a fixed K is fully determined by the s_C and s_R strategies. Now, following standard game theoretic practice, we define *optimal* strategies.

Definition 3. *For every graph G and $K \in \mathbb{N}$, the strategies $s_C^{(K)} \in \mathbf{S}_C^{(K)}$ and $s_R^{(K)} \in \mathbf{S}_R^{(K)}$ are a pair of optimal strategies if and only if*

$$\sup_{s_R \in \mathbf{S}_R^{(K)}} \inf_{s_C \in \mathbf{S}_C^{(K)}} T^{(K)}(s_C, s_R|G) = \inf_{s_C \in \mathbf{S}_C^{(K)}} \sup_{s_R \in \mathbf{S}_R^{(K)}} T^{(K)}(s_C, s_R|G) \tag{5}$$

The value *of the av-CR game played with K cops is the common value of the two sides of Equation (5) and we denote it $T^{(K)}\left(s_C^{(K)}, s_R^{(K)}|G\right)$.*

We emphasize that the validity of Equation (5) is not known *a priori*. C (resp. R) can guarantee that he loses no more than $\inf_{s_C \in \mathbf{S}_C^{(K)}} \sup_{s_R \in \mathbf{S}_R^{(K)}} T^{(K)}(s_C, s_R|G)$ (resp. gains no less than $\sup_{s_R \in \mathbf{S}_R^{(K)}} \inf_{s_C \in \mathbf{S}_C^{(K)}} T^{(K)}(s_C, s_R|G)$). We always have

$$\sup_{s_R \in \mathbf{S}_R^{(K)}} \inf_{s_C \in \mathbf{S}_C^{(K)}} T^{(K)}(s_C, s_R|G) \leq \inf_{s_C \in \mathbf{S}_C^{(K)}} \sup_{s_R \in \mathbf{S}_R^{(K)}} T^{(K)}(s_C, s_R|G) \tag{6}$$

But, since av-CR is an *infinite* game (*i.e.*, depending on s_C and s_R, it can last an infinite number of turns) it is not clear that equality holds in Equation (6) and, even when it does, the existence of optimal strategies $\left(s_C^{(K)}, s_R^{(K)}\right)$ which achieve the value is not guaranteed.

In fact it can be proved that, for $K \geq c(G)$, av-CR has both a value and optimal strategies. The details of this proof will be reported elsewhere, but the gist of the argument is the following. Since av-CR is played with $K \geq c(G)$ cops, by Theorem 2, C has a *memoryless* strategy which guarantees the game will last no more than $\overline{T}(K;G)$ turns. Hence av-CR with $K \geq c(G)$ essentially is a *finite* zero-sum two-player game; it is well known [35] that every such game has a value and optimal memoryless strategies. In short, we have the following.

Theorem 3. *Given any graph G and any $K \geq c(G)$, for the av-CR game there exists a pair* $\left(\sigma_C^{(K)}, \sigma_R^{(K)}\right) \in \widetilde{\mathbf{S}}_C^{(K)} \times \widetilde{\mathbf{S}}_R^{(K)}$ *of memoryless time optimal strategies such that*

$$T^{(K)}\left(\sigma_C^{(K)}, \sigma_R^{(K)}|G\right) = \sup_{s_R \in \mathbf{S}_R^{(K)}} \inf_{s_C \in \mathbf{S}_C^{(K)}} T^{(K)}(s_C, s_R|G) = \inf_{s_C \in \mathbf{S}_C^{(K)}} \sup_{s_R \in \mathbf{S}_R^{(K)}} T^{(K)}(s_C, s_R|G)$$

Hence we can define the capture time of a graph to be the value of av-CR when played on G with $K = c(G)$ cops.

Definition 4. *The* adversarial visible capture time *of G is*

$$ct(G) = \sup_{s_R \in \mathbf{S}_R^{(K)}} \inf_{s_C \in \mathbf{S}_C^{(K)}} T^{(K)}(s_C, s_R|G) = \inf_{s_C \in \mathbf{S}_C^{(K)}} \sup_{s_R \in \mathbf{S}_R^{(K)}} T^{(K)}(s_C, s_R|G)$$

with $K = c(G)$.

3.2. The Node dv-CR Game

In this game the robber is visible and performs a random walk on G (*drunk* robber) as indicated by Equation (1). In the absence of cops, Y_t is a Markov chain on V, with transition probability matrix P, where for every $u, v \in \{1, 2, ..., |V|\}$ we have

$$P_{u,v} = \Pr\left(Y_{t+1} = u | Y_t = v\right)$$

In the presence of one or more cops, $\{Y_t\}_{t=0}^{\infty}$ is a *Markov decision process* (MDP) [36] with state space $V \cup \{n+1\}$ (where $n+1$ is the *capture state*) and transition probability matrix $P(X_t)$ (obtained from P as shown in [37]); in other words, X_t is the *control* variable, selected by C.

Since no robber strategy is involved, the capture time on G only depends on the (K-cops strategy) s_C: namely:

$$T^{(K)}(s_C|G) = \min\left\{t : \exists k \in [K] \text{ such that } Y_t = X_t^k\right\} \tag{7}$$

which can also be written as

$$T^{(K)}(s_C|G) = \sum_{t=0}^{\infty} \mathbf{1}(Y_t \notin X_t) \tag{8}$$

where $\mathbf{1}(Y_t \notin X_t)$ equals 1 if Y_t does not belong to X_t (taken as a set of cop positions) and 0 otherwise. Since the robber performs a random walk on G, it follows that $T^{(K)}(s_C|G)$ is a random variable, and C wants to minimize its expected value:

$$E\left(T^{(K)}(s_C|G)\right) = E\left(\sum_{t=0}^{\infty} \mathbf{1}(Y_t \notin X_t)\right) \tag{9}$$

The minimization of Equation (9) is a typical *undiscounted, infinite horizon* MDP problem. Using standard MDP results [36] we see that (i) C loses nothing by determining $X_0, X_1, \ldots$ through a memoryless strategy $\sigma_C(x, y)$ and (ii) for every $K \geq 1$, $E\left(T^{(K)}(\sigma_C|G)\right)$ is well defined. Furthermore, for every $K \in \mathbb{N}$ there exists an optimal strategy $\sigma_C^{(K)}$ which minimizes $E\left(T^{(K)}(\sigma_C|G)\right)$; hence we have the following.

Theorem 4. *Given any graph G and $K \in \mathbb{N}$, for the dv-CR game played on G with K cops there exists a memoryless strategy $\sigma_C^{(K)} \in \widetilde{\mathbf{S}}_C^{(K)}$ such that*

$$E\left(T^{(K)}\left(\sigma_C^{(K)}|G\right)\right) = \inf_{s_C \in \mathbf{S}_C^{(K)}} E\left(T^{(K)}(s_C|G)\right)$$

Definition 5. *The* drunk visible capture time *of G is*

$$dct(G) = \inf_{s_C \in \mathbf{S}_C^{(K)}} E\left(T^{(K)}(s_C|G)\right)$$

with $K = c(G)$.

Note that, even though a single cop suffices to capture the drunk robber on any G, we have chosen to define $dct(G)$ to be the capture time for $K = c(G)$ cops; we have done this to make (in Section 4) an equitable comparison between $ct(G)$ and $dct(G)$.

3.3. The Node ai-CR Game

This is *not* a perfect information game, since C cannot see R's moves. Hence C and R must use *mixed* strategies s_C, s_R. A mixed strategy s_C (resp. s_R) specifies, for every t, a conditional probability $\Pr(X_t|X_0, Y_0, \ldots, Y_{t-2}, X_{t-1}, Y_{t-1})$ (resp. $\Pr(Y_t|X_0, Y_0, \ldots, Y_{t-1}, X_t)$) according to which C (resp. R) selects his t-th move. Let $\overline{\mathbf{S}}_C^{(K)}$ (resp. $\overline{\mathbf{S}}_R^{(K)}$) be the set of all mixed cop (resp. robber) strategies. A strategy pair $(s_R, s_C) \in \overline{\mathbf{S}}_C^{(K)} \times \overline{\mathbf{S}}_R^{(K)}$, specifies probabilities for all events $(X_0 = x_0, \ldots, X_t = x_t, Y_0 = y_0, \ldots, Y_t = y_t)$ and these induce a probability measure which in turn determines R's expected gain (and C's expected loss), namely $E\left(T^{(K)}\left(s_C^{(K)}, s_R^{(K)}|G\right)\right)$. Let us define

$$\underline{v}^{(K)} = \sup_{s_R \in \overline{\mathbf{S}}_R^{(K)}} \inf_{s_C \in \overline{\mathbf{S}}_C^{(K)}} E\left(T^{(K)}(s_C, s_R|G)\right)$$

$$\overline{v}^{(K)} = \inf_{s_C \in \overline{\mathbf{S}}_C^{(K)}} \sup_{s_R \in \overline{\mathbf{S}}_R^{(K)}} E\left(T^{(K)}(s_C, s_R|G)\right)$$

Similarly to av-CR, C (resp. R) can guarantee an expected payoff no greater than $\overline{v}^{(K)}$ (resp. no less than $\underline{v}^{(K)}$). If $\underline{v}^{(K)} = \overline{v}^{(K)}$, we denote the common value by $v^{(K)}$ and call it the *value* of the ai-CR game (played on G, with K cops). A pair of strategies $\left(s_C^{(K)}, s_R^{(K)}\right)$ is called optimal if and only if $E\left(T^{(K)}\left(s_C^{(K)}, s_R^{(K)}|G\right)\right) = v^{(K)}$.

In [9] we have studied the ai-CR game and proved that it does indeed have a value and optimal strategies. We give a summary of the relevant argument; proofs can be found in [9].

First, *invisibility does not increase the cop number*. In other words, there is a cop strategy (involving $c(G)$ cops) which guarantees bounded expected capture time for every robber strategy s_R. More precisely, we have proved the following.

Theorem 5. *On any graph G let $\overline{s}_C^{(K)}$ denote the strategy in which K cops random-walk on G. Then*

$$\forall K \geq c(G): \sup_{s_R \in \overline{\mathbf{S}}_R^{(K)}} E\left(T^{(K)}\left(\overline{s}_C^{(K)}, s_R | G\right)\right) < \infty$$

Now consider the "m-truncated" ai-CR game which is played exactly as the "regular" ai-CR but lasts at most m turns. Strategies $s_R \in \overline{\mathbf{S}}_R^{(K)}$ and $s_C \in \overline{\mathbf{S}}_C^{(K)}$ can be used in the m-truncated game: C and R use them only until the m-th turn. Let R receive one payoff unit for every turn in which the robber is not captured; denote the payoff of the m-truncated game (when strategies s_C, s_R are used) by $T_m^{(K)}(s_C, s_R | G)$. Clearly

$$\forall m \in \mathbb{N}, s_R \in \overline{\mathbf{S}}_R^{(K)}, s_C \in \overline{\mathbf{S}}_C^{(K)} : T_m^{(K)}(s_C, s_R | G) \leq T_{m+1}^{(K)}(s_C, s_R | G) \leq T^{(K)}(s_C, s_R | G)$$

The expected payoff of the m-truncated game is $E\left(T_m^{(K)}(s_C, s_R | G)\right)$. Because it is a *finite*, two-person, zero-sum game, the m-truncated game has a value and optimal strategies. Namely, the value is

$$v^{(K,m)} = \sup_{s_R \in \overline{\mathbf{S}}_R^{(K)}} \inf_{s_C \in \overline{\mathbf{S}}_C^{(K)}} E\left(T_m^{(K)}(s_C, s_R | G)\right) = \inf_{s_C \in \overline{\mathbf{S}}_C^{(K)}} \sup_{s_R \in \overline{\mathbf{S}}_R^{(K)}} E\left(T_m^{(K)}(s_C, s_R | G)\right)$$

and there exist optimal strategies $s_C^{(K,m)} \in \overline{\mathbf{S}}_C^{(K)}$, $s_R^{(K,m)} \in \overline{\mathbf{S}}_R^{(K)}$ such that

$$E\left(T_m^{(K)}\left(s_C^{(K,m)}, s_R^{(K,m)} | G\right)\right) = v^{(K,m)} < \infty \tag{10}$$

In [9] we use the truncated games to prove that the "regular" ai-CR game has a value, an optimal C strategy and ε-optimal R strategies. More precisely, we prove the following.

Theorem 6. *Given any graph G and $K \geq c(G)$, the ai-CR game played on G with K cops has a value $v^{(K)}$ which satisfies*

$$\lim_{m \to \infty} v^{(K,m)} = \underline{v}^{(K)} = \overline{v}^{(K)} = v^{(K)}$$

Furthermore, there exists a strategy $s_C^{(K)} \in \overline{\mathbf{S}}_C^{(K)}$ such that

$$\sup_{s_R \in \overline{\mathbf{S}}_R^{(K)}} E\left(T^{(K)}\left(s_C^{(K)}, s_R\right)\right) = v^{(K)} \tag{11}$$

and for every $\varepsilon > 0$ there exists an m_ε and a strategy $s_R^{(K,\varepsilon)}$ such that

$$\forall m \geq m_\varepsilon : v^{(K)} - \varepsilon \leq \sup_{s_C \in \overline{\mathbf{S}}_C^{(K)}} E\left(T^{(K)}\left(s_C, s_R^{(K,\varepsilon)}\right) | G\right) \leq v^{(K)} \tag{12}$$

Having established the existence of $v^{(K)}$ we have the following.

Definition 6. *The adversarial invisible capture time of G is*

$$ct_i(G) = v^{(K)} = \sup_{s_R \in \overline{\mathbf{S}}_R^{(K)}} \inf_{s_C \in \overline{\mathbf{S}}_C^{(K)}} E\left(T^{(K)}(s_C, s_R | G)\right) = \inf_{s_C \in \overline{\mathbf{S}}_C^{(K)}} \sup_{s_R \in \overline{\mathbf{S}}_R^{(K)}} E\left(T^{(K)}(s_C, s_R | G)\right)$$

with $K = c(G)$.

3.4. The Node di-CR Game

In this game Y_t is unobservable and drunk; call this the "regular" di-CR game and also introduce the m-truncated di-CR game. Both are one-player games or, equivalently, Y_t is a *partially observable MDP* (POMDP) [36]. The target function is

$$E\left(T^{(K)}\left(s_C|G\right)\right) = E\left(\sum_{t=0}^{\infty} \mathbf{1}\left(Y_t \notin X_t\right)\right) \tag{13}$$

which is exactly the same as Equation (9) but now Y_t is *unobservable*. Equation (13) can be approximated by

$$E\left(T_m^{(K)}\left(s_C|G\right)\right) = E\left(\sum_{t=0}^{m} \mathbf{1}\left(Y_t \notin X_t\right)\right) \tag{14}$$

The expected values in Equations (13) and (14) are well defined for every s_C. C must select a strategy $s_C \in \overline{\mathbf{S}}_C^{(K)}$ which minimizes $E\left(T^{(K)}\left(s_C|G\right)\right)$. This is a typical *infinite horizon, undiscounted* POMDP problem [36] for which the following holds.

Theorem 7. *Given any graph G and $K \in \mathbb{N}$, for the di-CR game played on G with K cops there exists a strategy $s_C^{(K)} \in \overline{\mathbf{S}}_C^{(K)}$ such that*

$$E\left(T^{(K)}\left(s_C^{(K)}|G\right)\right) = \inf_{s_C \in \overline{\mathbf{S}}_C^{(K)}} E\left(T^{(K)}\left(s_C|G\right)\right)$$

Hence we can introduce the following.

Definition 7. *The* drunk invisible capture time *of G is*

$$dct_i\left(G\right) = \inf_{s_C \in \overline{\mathbf{S}}_C^{(K)}} E\left(T^{(K)}\left(s_C|G\right)\right)$$

with $K = c\left(G\right)$.

3.5. The Edge CR Games

As already mentioned, every edge CR variant can be reduced to the corresponding node variant played on $L\left(G\right)$, the line graph of G. Hence all the results and definitions of Sections 3.1–3.4 hold for the edge variants as well. In particular, we have an edge cop number $\overline{c}\left(G\right) = c\left(L\left(G\right)\right)$ and capture times

$$\overline{ct}\left(G\right) = ct\left(L\left(G\right)\right), \quad \overline{dct}\left(G\right) = dct\left(L\left(G\right)\right), \quad \overline{ct}_i\left(G\right) = ct_i\left(L\left(G\right)\right), \quad \overline{dct}_i\left(G\right) = dct_i\left(L\left(G\right)\right)$$

In general, all of these "edge CR parameters" will differ from the corresponding "node CR parameters".

4. The Cost of Visibility

4.1. Cost of Visibility in the Node CR Games

As already remarked, we expect that ai-CR is more difficult (from C's point of view) than av-CR (the same holds for the drunk counterparts of this game). We quantify this statement by introducing the *cost of visibility* (COV).

Definition 8. *For every G, the* adversarial cost of visibility *is $H_a(G) = \frac{ct_i(G)}{ct(G)}$ and the* drunk cost of visibility *is $H_d(G) = \frac{dct_i(G)}{dct(G)}$.*

Clearly, for every G we have $H_a(G) \geq 1$ and $H_d(G) \geq 1$ (*i.e.*, it is at least as hard to capture an invisible robber than a visible one). The following theorem shows that in fact both $H_a(G)$ and $H_d(G)$ can become arbitrarily large. In proving the corresponding theorems, we will need the family of *long star graphs* $S_{N,M}$. For specific values of M and N, $S_{N,M}$ consists of N paths (we call these *rays*) each having M nodes, joined at a central node, as shown in Figure 1.

Figure 1. (a): the *star* graph $S_{N,1}$; **(b)**: the *long star* graph $S_{N,M}$.

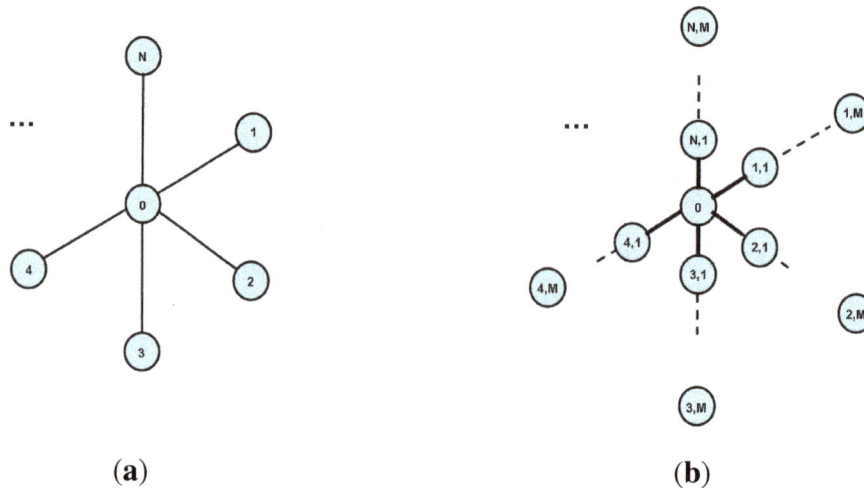

(a)   (b)

Theorem 8. *For every $N \in \mathbb{N}$ we have $H_a(S_{N,1}) = N$.*

Proof. (i) Computing $ct(S_{N,1})$. In av-CR, for every $N \in \mathbb{N}$ we have $ct(S_{N,1}) = 1$: the cop starts at $X_0 = 0$, the robber starts at some $Y_0 = u \neq 0$ and, at $t = 1$, he is captured by the cop moving into u; *i.e.*, $ct(S_{N,1}) \leq 1$; on the other hand, since there are at least two vertices ($N \geq 1$), clearly $ct(S_{N,1}) \geq 1$. **(ii) Computing $ct_i(S_{N,1})$.** Let us now show that in ai-CR we have $ct_i(S_{N,1}) = N$. C places the cop at $X_0 = 0$ and R places the robber at some $Y_0 = u \neq 0$. We will obtain $ct_i(S_{N,1})$ by bounding it from above and below. For an upper bound, consider the following C strategy. Since C does not know the robber's location, he must check the leaf nodes one by one. So at every odd t he moves the cop into some $u \in \{1, 2, \ldots, N\}$ and at every even t he returns to 0. Note that R cannot change the robber's original position; in order to do this, the robber must pass through 0 but then he will be captured by the cop (who either is already in 0 or will be moved into it just after the robber's move). Hence C can choose the nodes he will check on odd turns with uniform probability and *without repetitions*. Equivalently, we can assume that the order in which nodes are chosen by C is selected uniformly at random from the set of all permutations; further, we assume that R (who does not know this order) starts at some $Y_0 = u \in \{1, \ldots, N\}$. Then we have

$$ct_i\left(S_{N,1}\right) \le \frac{1}{N} \cdot 1 + \frac{1}{N} \cdot 3 + \ldots + \frac{1}{N} \cdot (2N - 1) = N$$

For a lower bound, consider the following R strategy. The robber is initially placed at a random leaf that is different than the one selected by C (if the cop did not start at the center). Knowing this, the best C strategy is to check (in any order) all leaves without repetition. If the cop starts at the center, we get exactly the same sum as for the upper bound. Otherwise, we have

$$ct_i\left(S_{N,1}\right) \ge \frac{1}{N-1} \cdot 2 + \frac{1}{N-1} \cdot 4 + \ldots + \frac{1}{N-1} \cdot (2N - 2) = N$$

(iii) Computing $H_a\left(S_{N,1}\right)$. Hence, for all $N \in \mathbb{N}$ we have $H_a\left(S_{N,1}\right) = \frac{ct_i\left(S_{N,1}\right)}{ct\left(S_{N,1}\right)} = N$ $\quad\square$

Theorem 9. *For every $N \in \mathbb{N} - \{1\}$ we have*

$$H_d\left(S_{N,M}\right) = (1 + o(1))\frac{(2N - 1)(N - 1) + 1}{N} \ge 2N - 3$$

where the asymptotics is with respect to M; N is considered a fixed constant.

Proof. (i) Computing $dct\left(S_{N,M}\right)$. We will first show that, for any $N \in \mathbb{N}$, we have $dct\left(S_{N,M}\right) = (1 + o(1))\frac{M}{2}$ (recall that the parameter N is a fixed constant whereas $M \to \infty$.) Suppose that the cop starts on the i-th ray, at distance $(1 + o(1))cM$ from the center (for some constant $c \in [0, 1]$). The robber starts at a random vertex. It follows that for any j such that $1 \le j \le N$, the robber starts on the j-th ray with probability $(1 + o(1))/N$. It is a straightforward application of Chernoff bounds to show that with probability $1 + o(1)$ the robber will not move by more than $o(M)$ in the next $O(MN) = O(M)$ steps, which suffice to finish the game. This is so because, if X has a binomial distribution $Bin(n, p)$, then $Pr(|X - np| \ge \epsilon np) \le 2\exp(-\epsilon^2 np/3)$ for any $\epsilon \le 3/2$. Now suppose the robber starts at distance $\omega(M^{2/3})$ from the center. During $N = O(M)$ steps the robber makes in expectation $N/2$ steps towards the center, and $N/2$ steps towards the end of the ray. The probability to make during N steps more than $N/2 + M^{2/3}$ steps towards the center, say, is thus at most $e^{-cM^{1/3}}$, and the same holds also by taking a union bound over all $O(M)$ steps. Hence, with probability at least $1 - e^{-cM^{1/3}}$ he will throughout $O(M)$ steps remain at distance $O(M^{2/3})$ from his initial position. In short, the expected capture time is easy to calculate.

- With probability $(1 - c + o(1))/N$, the robber starts on the same ray as the cop but farther away from the center. Conditioning on this event, the expected capture time is $M(1 - c + o(1))/2$.
- With probability $(c + o(1))/N$, the robber starts on the same ray as the cop but closer to the center. Conditioning on this event, the expected capture time is $M(c + o(1))/2$.
- With probability $(N - 1 + o(1))/N$, the robber starts on different ray than the cop. Conditioning on this event, the expected capture time is $(c + o(1))M + M(1/2 + o(1))$.

It follows that the expected capture time is

$$(1 + o(1))M \left(\frac{1-c}{N} \cdot \frac{1-c}{2} + \frac{c}{N} \cdot \frac{c}{2} + \frac{N-1}{N} \cdot \frac{2c+1}{2} \right)$$

which is maximized for $c = 0$, giving $dct\,(S_{N,M}) = (1 + o\,(1)) \frac{M}{2}$.

(ii) Computing $dct_i\,(S_{N,M})$. The initial placement for the robber is the same as in the visible variant, that is, the uniform distribution is used. However, since the robber is now invisible, C has to check all rays. As before, by Chernoff bounds, with probability at least $1 - e^{-cM^{1/3}}$ (for some constant $c > 0$) during $O(M)$ steps the robber is always within distance $O(M^{2/3})$ from its initial position. If the robber starts at distance $\omega(M^{2/3})$ from the center, he will thus with probability at least $1 - e^{-cM^{1/3}}$ not change his ray during $O(M)$ steps. Otherwise, he might change from one ray to the other with bigger probability, but note that this happens only with the probability of the robber starting at distance $O(M^{2/3})$ from the center, and thus with probability at most $O(M^{-1/3})$. Keeping these remarks in mind, let us examine "reasonable" C strategies. It turns out there exist three such.

(ii.1) Suppose C starts at the end of one ray (chosen arbitrarily), goes to the center, and then successively checks the remaining rays without repetition, with probability at least $1 - O(M^{-1/3})$, the robber will be caught. If the robber is caught (this implies that the robber did not switch rays), the capture time is calculated as follows:

- With probability $(1 + o(1))/N$, the robber starts on the same ray as the cop. Conditioning on this event, the expected capture time is $(1 + o(1))M/2$.
- With probability $(1 + o(1))/N$, the robber starts on the j-th ray visited by the cop. Conditioning on this event, the expected capture time is $(1 + o(1))(M + 2M(j-2) + M/2)$. ($M$ steps are required to move from the end of the first ray to the center, $2M$ steps are 'wasted' to check $j - 2$ rays, and $M/2$ steps are needed to catch the robber on the j-th ray, on expectation.)

Hence, conditioned under not switching rays, the expected capture time in this case is

$$(1 + o(1)) \frac{M}{N} \left(\frac{1}{2} + \left(1 + \frac{1}{2} \right) + \left(3 + \frac{1}{2} \right) + \ldots + \left(1 + 2(N-2) + \frac{1}{2} \right) \right)$$

$$= (1 + o(1)) \frac{M}{N} \left(\frac{1}{2} + \left(2 \cdot 1 - \frac{1}{2} \right) + \left(2 \cdot 2 - \frac{1}{2} \right) + \ldots + \left(2(N-1) - \frac{1}{2} \right) \right)$$

$$= (1 + o(1)) \frac{M}{N} \left(\frac{1}{2} + \frac{2N-1}{2} \cdot (N-1) \right)$$

$$= (1 + o(1)) \frac{M}{2} \cdot \frac{(2N-1)(N-1)+1}{N}$$

Otherwise, if the robber is not caught, C just randomly checks rays: starting from the center, C chooses a random ray, goes until the end of the ray, returns to the center, and continues like this, until the robber is caught. The expected capture time in this case is

$$\sum_{j \geq 1} \left((1 - \frac{1}{N})^{j-1} \frac{1}{N} \left(2(j-1)M + M/2 \right) \right) = O(MN) = O(M)$$

Since this happens with probability $O(M^{-1/3})$, the contribution of the case where the robber switches rays is $o(M)$, and therefore for this strategy of C, the expected capture time is

$$(1 + o(1)) \frac{M}{2} \cdot \frac{(2N-1)(N-1)+1}{N}$$

(ii.2) Now suppose C starts at the center of the ray, rather than the end, and checks all rays from there. By the same arguments as before, the capture time is

$$(1 + o(1))\frac{M}{N}\left(\frac{1}{2} + \left(2 + \frac{1}{2}\right) + \left(4 + \frac{1}{2}\right) + \ldots + \left(2 + 2(N-2) + \frac{1}{2}\right)\right)$$

which is worse than in the case when starting at the end of a ray.

(ii.3) Similarly, suppose the cop starts at distance cM from the center, for some $c \in [0, 1]$. If he first goes to the center of the ray, and then checks all rays (suppose the one he came from is the last to be checked), then the capture time is

$$(1 + o(1))\frac{M}{N}\left(\frac{c^2}{2} + \left(c + \frac{1}{2}\right) + \left(c + 2 + \frac{1}{2}\right) + \ldots + \right.$$
$$\left.\left(\left(c + 2(N-2) + \frac{1}{2}\right) + (1-c)\left(2c + 2(N-1) + \frac{1-c}{2}\right)\right)\right)$$

which is minimized for $c = 1$. And if C goes first to the end of the ray, and then to the center, the capture time is

$$(1 + o(1))\frac{M}{N}\left(\frac{((1-c)^2}{2} + c\left(2(1-c) + \frac{c}{2}\right) + \left(2(1-c) + c + \frac{1}{2}\right) + \ldots + \right.$$
$$\left.\left(\left(2(1-c) + c + 2(N-2) + \frac{1}{2}\right)\right)\right)$$

which for $N \geq 2$ is also minimized for $c = 1$ (in fact, for $N = 2$ the numbers are equal).

In short, the smallest capture time is achieved when C starts at the end of some ray and therefore

$$dct_i(S_{N,M}) = (1 + o(1))\frac{M}{2} \cdot \frac{(2N-1)(N-1)+1}{N}$$

(iii) Computing $H_d(S_{N,M})$. It follows that for all $N \in \mathbb{N} - \{1\}$ we have

$$H_d(S_{N,M}) = \frac{dct_i(S_{N,M})}{dct(S_{N,M})} = (1 + o(1))\frac{(2N-1)(N-1)+1}{N} \geq 2N - 3$$

completing the proof. $\square$

4.2. Cost of Visibility in the Edge CR Games

The cost of visibility in the edge CR games is defined analogously to that of node games.

Definition 9. *For every G, the* edge adversarial cost of visibility *is* $\overline{H}_a(G) = \frac{\overline{ct_i}(G)}{\overline{ct}(G)}$ *and the* edge drunk cost of visibility *is defined as* $\overline{H}_d(G) = \frac{\overline{dct_i}(G)}{\overline{dct}(G)}$.

Clearly, for every G we have $\overline{H}_a(G) \geq 1$ and $\overline{H}_d(G) \geq 1$. The following theorems show that in fact both $\overline{H}_a(G)$ and $\overline{H}_d(G)$ can become arbitrarily large. To prove these theorems we will use the previously introduced star graph $S_{N,1}$ and its line graph which is the clique K_N. These graphs are illustrated in Figure 2 for $N = 6$.

Figure 2. (a): the *star* graph $S_{6,1}$ and **(b):** its line graph, the clique K_6.

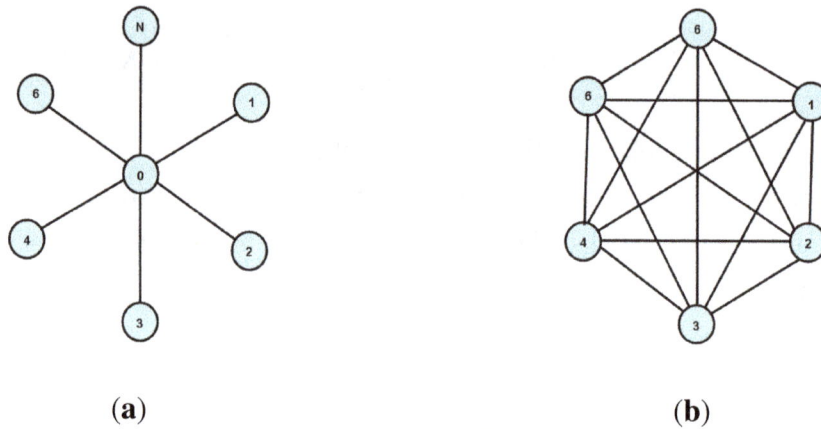

(a)   (b)

Theorem 10. *For every* $N \in \mathbb{N} - \{1\}$ *we have* $\overline{H}_a(S_{N,1}) = N - 1$.

Proof. We have $\overline{H}_a(S_{N,1}) = \frac{\overline{ct}_i(S_{N,1})}{\overline{ct}(S_{N,1})} = \frac{ct_i(K_N)}{ct(K_N)}$ and, since $N \geq 2$, clearly $ct(K_N) = 1$. Let us now compute $ct_i(K_N)$.

For an upper bound on $ct_i(K_N)$, C might just move to a random vertex. If the robber stays still or if he moves to a vertex different from the one occupied by C, he will be caught in the next step with probability $1/(N-1)$, and thus an upper bound on the capture time is $N - 1$.

For a lower bound, suppose that the robber always moves to a randomly chosen vertex, different from the one occupied by C, and including the one occupied by him now (that is, with probability $1/(N-1)$ he stands still, and after his turn, he is with probability $1/(N-1)$ at each vertex different from the vertex occupied by C. Hence C is forced to move, and since he has no idea where to go, the best strategy is also to move randomly, and the robber will be caught with probability $1/(N-1)$, yielding a lower bound on the capture time of $N - 1$. Therefore

$$ct_i(K_N) = N - 1$$

Hence

$$\overline{H}_a(S_{N,1}) = \frac{\overline{ct}_i(S_{N,1})}{\overline{ct}(S_{N,1})} = \frac{ct_i(K_N)}{ct(K_N)} = N - 1$$

□

Theorem 11. *For every* $N \in \mathbb{N} - \{1\}$ *we have* $\overline{H}_d(S_{N,1}) = \frac{N(N-1)}{2N-3}$.

Proof. This is quite similar to the adversarial case. We have $\overline{H}_d(S_{N,1}) = \frac{\overline{dct}_i(S_{N,1})}{\overline{dct}(S_{N,1})} = \frac{dct_i(K_N)}{dct(K_N)}$. Clearly we have $dct(K_N) = 1 - 1/N$ (with probability $1/N$ the robber selects the same vertex to start with as the cop and is caught before the game actually starts; otherwise is caught in the first round).

For $dct_i(K_N)$, it is clear that the strategy of constantly moving is best for the cop, as in this case there are two chances to catch the robber (either by moving towards him, or by afterwards the robber moving onto the cop). It does not matter where he moves to as long as he keeps moving, and we may thus assume that he starts at some vertex v and moves to some other vertex w in the first round, then comes back to v and oscillates like that until the end of the game. When the cop moves to another vertex, the probability that the robber is there is $1/(N-1)$. If he is still not caught, the robber moves to a random place, thereby selecting the vertex occupied by the cop with probability $1/(N-1)$. Hence, the probability to catch the robber in one step is $\frac{1}{N-1} + (1 - \frac{1}{N-1})\frac{1}{N-1} = \frac{2N-3}{(N-1)^2}$. Thus, this time the capture time is a geometric random variable with probability of success equal to $\frac{2N-3}{(N-1)^2}$. We get $dct_i(K_N) = \frac{(N-1)^2}{2N-3}$ and so

$$\overline{H}_d(S_{N,1}) = \frac{\overline{dct_i}(S_{N,1})}{\overline{dct}(S_{N,1})} = \frac{dct_i(K_N)}{dct(K_N)} = \frac{(N-1)^2/(2N-3)}{(N-1)/N} = \frac{N(N-1)}{2N-3}$$

which can become arbitrarily large by appropriate choice of N. $\quad\square$

5. Algorithms for COV Computation

For graphs of relatively simple structure (e.g., paths, cycles, full trees, grids) capture times and optimal strategies can be found by analytical arguments [9,37]. For more complicated graphs, an algorithmic solution becomes necessary. In this section we present algorithms for the computation of capture time in the previously introduced node CR variants. The same algorithms can be applied to the edge variants by replacing G with $L(G)$.

5.1. Algorithms for Visible Robbers

5.1.1. Algorithm for Adversarial Robber

The av-CR capture time $ct(G)$ can be computed in polynomial time. In fact, stronger results have been presented by Hahn and MacGillivray; in [31] they present an algorithm which, given K, computes for every $(x,y) \in V^2$ the following:

1. $C(x,y)$, the optimal game duration when the cop/robber configuration is (x,y) and it is C's turn to play;
2. $R(x,y)$, the optimal game duration when the cop/robber configuration is (x,y) and it is R's turn to play.

Note that, when $K < c(G)$, there exist (x,y) such that $C(x,y) = R(x,y) = \infty$; Hahn and MacGillivray's algorithm computes this correctly, as well.

The av-CR capture time can be computed by $ct(G) = \min_{x \in V} \max_{y \in V} C(x,y)$; the optimal search strategies $\hat{\sigma}_C, \hat{\sigma}_R$ can also be easily obtained from the *optimality equations*, as will be seen a little later. We have presented in [37] an implementation of Hahn and MacGillivray's algorithm, which we call CAAR (Cops Against Adversarial Robber). Below we present this, as Algorithm 1, for the case of a single cop (the generalization for more than one cop is straightforward).

The algorithm operates as follows. In lines `01-08` $C^{(0)}(x,y)$ and $R^{(0)}(x,y)$ are initialized to ∞, except for "diagonal" positions $(x,y) \in V_D^2$ (*i.e.*, positions with $x = y$) for which we obviously have $C(x,x) = R(x,x) = 0$. Then a loop is entered (lines `10-19`) in which $C^{(i)}(x,y)$ is computed (line `12`) by letting the cop move to the position which achieves the smallest capture time (according to the currently available estimate $R^{(i-1)}(x,y)$); $R^{(i)}(x,y)$ is computed similarly in line `13`, looking for the largest capture time. This process is repeated until no further changes take place, at which point the algorithm exits the loop and terminates. This algorithm is a game theoretic version of *value iteration* [36], which we see again in Section 5.2. It has been proved in [31] that, for any graph G and any $K \in \mathbb{N}$, CAAR always terminates and the finally obtained (C, R) pair satisfies the *optimality equations*

$$\forall (x,y) \in V_D^2 : C(x,y) = 0; \quad \forall (x,y) \in V^2 - V_D^2 : C(x,y) = 1 + \min_{x' \in N[x]} R(x',y) \qquad (15)$$

$$\forall (x,y) \in V_D^2 : R(x,y) = 0; \quad \forall (x,y) \in V^2 - V_D^2 : R(x,y) = 1 + \max_{y' \in N[y]} C(x,y') \qquad (16)$$

The optimal memoryless strategies $\sigma_C^{(K)}(x,y)$, $\sigma_R^{(K)}(x,y)$ can be computed for every position (x,y) by letting $\sigma_C^{(K)}(x,y)$ (resp. $\sigma_R^{(K)}(x,y)$) be a node $x' \in N[x]$ (resp. $y' \in N[y]$) which achieves the minimum in Equation (15) (resp. maximum in Equation (16)). The capture time $ct(G)$ is computed from

$$ct(G) = \min_{x \in V} \max_{y \in V} C(x,y)$$

Algorithm 1: Cops Against Adversarial Robber (CAAR)

Input: $G = (V, E)$

```
01    For All (x,y) ∈ V_D²
02        C^(0)(x,y) = 0
03        R^(0)(x,y) = 0
04    EndFor
05    For All (x,y) ∈ V² − V_D²
06        C^(0)(x,y) = ∞
07        R^(0)(x,y) = ∞
08    EndFor
09    i = 1
10    While 1 > 0
11        For All (x,y) ∈ V² − V_D²
12            C^(i)(x,y) = 1 + min_{x'∈N[x]} R^(i−1)(x',y)
13            R^(i)(x,y) = 1 + max_{y'∈N[y]} C^(i)(x,y')
14        EndFor
15        If C^(i) = C^(i−1) And R^(i) = R^(i−1)
16            Break
17        EndIf
18        i ← i+1
19    EndWhile
20    C = C^(i)
21    R = R^(i)
```

Output: C, R

5.1.2. Algorithm for Drunk Robber

For any given K, *value iteration* can be used to determine both $dct(G, K)$ and the optimal strategy $\sigma_C^{(K)}(x,y)$; one implementation is our CADR (Cops Against Drunk Robber) algorithm [37] which is a typical value-iteration [36] MDP algorithm; alternatively, CADR can be seen as an extension of the CAAR idea to the dv-CR. Below we present this, as Algorithm 2, for the case of a single cop (the generalization for more than one cops is straightforward).

Algorithm 2: Cops Against Drunk Robber (CADR)

Input: $G = (V, E), \varepsilon$

```
01 For All (x,y) ∈ V²_D
02   C⁽⁰⁾(x,y) = 0
03 EndFor
04 For All (x,y) ∈ V − V²_D
05   C⁽⁰⁾(x,y) = ∞
06 EndFor
07 i = 1
08 While 1 > 0
09   For All (x,y) ∈ V − V²_D
10     C⁽ⁱ⁾(x,y) = 1 + min_{x'∈N[x]} Σ_{y'∈V} P((x',y) → (x',y')) C⁽ⁱ⁻¹⁾(x',y')
11   EndFor
12   If max_{(x,y)∈V²} |C⁽ⁱ⁾(x,y) − C⁽ⁱ⁻¹⁾(x,y)| < ε
13       Break
14   EndIf
15   i ← i+1
16 EndWhile
17 C = C⁽ⁱ⁾
```

Output: C

The algorithm operates as follows (again we use $C(x,y)$ to denote the optimal expected game duration when the game position is (x,y)). In lines `01-06` $C^{(0)}(x,y)$ is initialized to ∞, except for "diagonal" positions $(x,y) \in V_D^2$. In the main loop (lines `08-16`) $C^{(i)}(x,y)$ is computed (line `10`) by letting the cop move to the position which achieves the smallest expected capture time $(P((x',y) \to (x',y'))$ in line `10` indicates the transition probability from (x',y) to $(x',y'))$. This process is repeated until the maximum change $|C^{(i)}(x,y) - C^{(i-1)}(x,y)|$ is smaller than the *termination criterion* ε, at which point the algorithm exits the loop and terminates. This is a typical *value iteration* MDP algorithm [36]; the convergence of such algorithms has been studied by several authors, in various degrees of generality [38–40]. A simple yet strong result, derived in [39], uses the concept of *proper strategy*: a strategy is called proper if it yields finite expected capture time. It is proved in [39] that, if a proper strategy exists for graph G, then CADR-like algorithms converge. In the case of dv-CR we know that C has a proper strategy: it is the random walking strategy $\bar{s}_C^{(K)}$ mentioned in Theorem 5. Hence CADR converges and in the limit, $C = \lim_{i\to\infty} C^{(i)}$ satisfies the *optimality equations*

$$\forall (x,y) \in V_D^2 : C(x,y) = 0; \quad \forall (x,y) \in V^2 - V_D^2 : C(x,y) = 1 + \min_{x'\in N[x]} \sum P((x',y) \to (x',y')) C(x',y')$$

$$(17)$$

The optimal memoryless strategy $\sigma_C^{(K)}(x, y)$ can be computed for every position (x, y) by letting $\sigma_C^{(K)}(x, y)$ be a node $x' \in N[x]$ (resp. $y' \in N[y]$) which achieves the minimum in Equation (15) (resp. maximum in Equation (16)). The capture time $dct(G)$ is computed from

$$dct(G) = \min_{x \in V} C(x, y)$$

5.2. Algorithms for Invisible Robbers

5.2.1. Algorithms for Adversarial Robber

We have not been able to find an efficient algorithm for solving the ai-CR game. Several algorithms for *imperfect information stochastic games* could be used to this end but we have found that they are practical only for very small graphs. The problem is that for every game position (e.g., assuming one robber and one cop, for a triple (x, y, p) indicating cop-position, robber-position and player to move) a full two-player, one-turn sub-game must be solved; this must be done for $2 \cdot |V|^2$ positions and for sufficient iterations to achieve convergence. The computational load quickly becomes unmanageable.

5.2.2. Algorithm for Drunk Robber

In the case of the drunk invisible robber we are also using a game tree search algorithm with pruning, for which some analytical justification can be provided. We call this the *Pruned Cop Search* (PCS) algorithm. Before presenting the algorithm we will introduce some notation and then prove a simple fact about expected capture time. We limit ourselves to the single cop case, since the extension to more cops is straightforward.

We let $\mathbf{x} = x_0 x_1 x_2 \ldots$ be an infinite history of cop moves. Letting t being the current time step, the probability vector $\mathbf{p}(t)$ contains the probabilities of the robber being in node $v \in V$ or in the *capture state* $n + 1$; more specifically: $\mathbf{p}(t) = [p_1(t), \ldots, p_v(t), \ldots, p_n(t), p_{n+1}(t)]$ and $p_v(t) = \Pr(y_t = v | x_0 x_1 \ldots x_t)$. Hence $\mathbf{p}(t)$ depends (as expected) on the *finite* cop history $x_0 x_1 \ldots x_t$. The expected capture time is denoted by $C(\mathbf{x}) = E(T|\mathbf{x})$; the conditioning is on the infinite cop history. The PCS algorithm works because $E(T|\mathbf{x})$ can be approximated from a finite part of $\mathbf{x}$, as explained below. We have

$$C(\mathbf{x}) = E(T|\mathbf{x}) = \sum_{t=0}^{\infty} t \cdot \Pr(T = t|\mathbf{x}) = \sum_{t=0}^{\infty} \Pr(T > t|\mathbf{x}) \qquad (18)$$

$\mathbf{x}$ in the conditioning is the *infinite* history $\mathbf{x} = x_0 x_1 x_2 \ldots$. However, for every t we have

$$\Pr(T > t|\mathbf{x}) = 1 - \Pr(T \le t|\mathbf{x}) = 1 - \Pr(T \le t|x_0 x_1 \ldots x_t)$$

Let us define

$$C^{(t)}(x_0 x_1 \ldots x_t) = \sum_{\tau=0}^{t} [1 - \Pr(T \le \tau|x_0 x_1 \ldots x_\tau)] = \sum_{\tau=0}^{t} [1 - p_{n+1}(\tau)]$$

where $p_{n+1}(\tau)$ is the probability that the robber is in the *capture state* $n + 1$ at time τ (the dependence on $x_0 x_1 \ldots x_\tau$ is suppressed, for simplicity of notation). Then for all t we have

$$C^{(t)}(x_0 x_1 \ldots x_t) = C^{(t-1)}(x_0 x_1 \ldots x_{t-1}) + (1 - p_{n+1}(t)) \qquad (19)$$

Update Equation (19) can be computed using only the previous cost $C^{(t-1)} (x_0 x_1 \ldots x_{\tau-1})$ and the (previously computed) probability vector $\mathbf{p}(t)$. While $C^{(t)} (x_0 \ldots x_t) \leq C(\mathbf{x})$, we hope that (at least for the "good" histories) we have

$$\lim_{t \to \infty} C^{(t)} (x_0 \ldots x_t) = C(\mathbf{x}) \tag{20}$$

This approximation works well, with $C^{(t)} (x_0 \ldots x_t)$ approaching its limiting value when t is in the range 15 to 20.

Below we present this, as Algorithm 3, in pseudocode. We have introduced a structure S with fields $S.\mathbf{x}$, $S.\mathbf{p}$, $S.C = C(S.\mathbf{x})$. Also we denote concatenation by the & symbol, i.e., $x_0 x_1 \ldots x_t \& v = x_0 x_1 \ldots x_t v$.

Algorithm 3: Pruned Cop Search (PCS)

Input: $G = (V, E)$, x_0, J_{max}, ε

```
01   t = 0
02   S.x = x₀, S.p = Pr(y₀|x₀), S.C = 0
03   S = {S}
04   C_best^old = 0
05   While 1 > 0
06         S̃ = ∅
07         For All S ∈ S
08               x = S.x,  p = S.p,  C = S.C
09               For All v ∈ N[xₜ]
10                     x' = x&v
11                     p' = p · P(v)
12                     C' = Cost(x', p', C)
13                     S'.x = x',  S'.p = p',  S'.C = C'
14                     S̃ = S̃ ∪ {S'}
15               EndFor
16         EndFor
17         S = Prune(S̃, J_max)
18         [x_best, C_best] = Best(S)
19         If |C_best − C_best^old| < ε
20               Break
21         Else
22               C_best^old = C_best
23               t ← t + 1
24         EndIf
25   EndWhile
```

Output: $\mathbf{x}_{best}$, $C_{best} = C(\mathbf{x}_{best})$.

The PCS algorithm operates as follows. At initialization (lines 01–04), we create a single S structure (with $S.\mathbf{x}$ being the initial cop position, $S.\mathbf{p}$ the initial, uniform robber probability and $S.C = 0$) which we store in the set **S**. Then we enter the main loop (lines 05–25) where we pick each available cop sequence **x** of length t (line 08). Then, in lines 09–15 we compute, for all legal extensions $\mathbf{x}' = \mathbf{x} \& v$ (where $v \in N[x_t]$) of length $t + 1$ (line 10), the corresponding $\mathbf{p}'$ (line 11) and C' (by the subroutine **Cost** at line 12). We store these quantities in S' which is placed in the temporary storage set $\widetilde{\mathbf{S}}$ (lines 13–14). After exhausting all possible extensions of length $t + 1$, we prune the temporary set $\widetilde{\mathbf{S}}$, retaining only the $J_{\max}$ best cop sequences (this is done in line 17 by the subroutine **Prune** which computes "best" in terms of smallest $C(\mathbf{x})$). Finally, the subroutine **Best** in line 18 computes the overall smallest expected capture time $C_{best} = C(\mathbf{x}_{best})$. The procedure is repeated until the termination criterion $|C_{best} - C_{best}^{old}| < \varepsilon$ is satisfied. As explained above, the criterion is expected to be always eventually satisfied because of Equation (20).

6. Experimental Estimation of the Cost of Visibility

We now present numerical computations of the drunk cost of visibility for graphs which are not amenable to analytical computation. We do not deal with the adversarial cost of visibility because, while we can compute $ct(G)$ with the CAAR algorithm, we do not have an efficient algorithm to compute $ct_i(G)$; hence we cannot perform experiments on $H_a(G) = \frac{ct_i(G)}{ct(G)}$. The difficulty with $ct_i(G)$ is that ai-CR is a stochastic game of imperfect information; even for very small graphs, one cop and one robber, ai-CR involves a state space with size far beyond the capabilities of currently available stochastic games algorithms (see [41]). In Section 6.1 we deal with node games and in Section 6.2 with edge games.

6.1. Experiments with Node Games

Since $H_d(G) = \frac{dct_i(G)}{dct(G)}$, we use the CADR algorithm to compute $dct(G)$ and the PCS algorithm to compute $dct_i(G)$. We use graphs G obtained from *indoor environments*, which we represent by their *floorplans*. In Figure 3 we present a floorplan and its graph representation. The graph is obtained by decomposing the floorplan into convex *cells*, assigning each cell to a node and connecting nodes by edges whenever the corresponding cells are connected by an open space.

Figure 3. A floorplan and the corresponding graph.

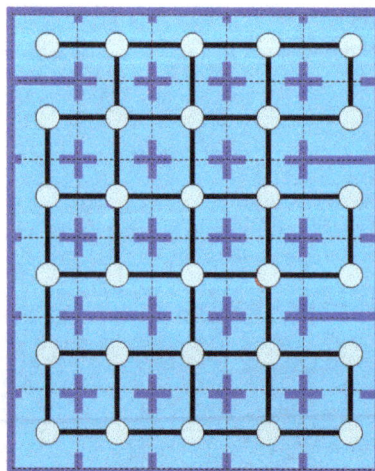

We have written a script which, given some parameters, generates random floorplans and their graphs. Every floorplan consists of a rectangle divided into orthogonal "rooms". If each internal room were connected to its four nearest neighbors we would get an $M \times N$ grid G'. However, we randomly generate a spanning tree G_T of G' and initially introduce doors only between rooms which are connected in G_T. Our final graph G is obtained from G_T by iterating over all missing edges and adding each one with probability $p_0 \in [0, 1]$. Hence each floorplan is characterized by three parameters: M, N and p_0.

We use the following pairs of (M, N) values: (1,30), (2,15), (3,10), (4,7), (5,6). Four of these pairs give a total of 30 nodes and the pair ($M = 4$, $N = 7$) gives $n = 28$ nodes; as M/N increases, we progress from a path to a nearly square grid. For each (M, N) pair we use five p_0 values: 0.00, 0.25, 0.50, 0.75, 1.00; note the progression from a tree ($p_0 = 0.00$) to a full grid ($p_0 = 1.00$). For each triple (M, N, p_0) we generate 50 floorplans, obtain their graphs and for each graph G we compute $dct(G)$ using CADR, $dct_i(G)$ using PCS and $H_d(G) = \frac{dct_i(G)}{dct(G)}$; finally we average $H_d(G)$ over the 50 graphs. In Figure 4 we plot $dct(G)$ as a function of the probability p_0; each plotted curve corresponds to an (M, N) pair. Similarly, in Figure 5 we plot $dct_i(G)$ and in Figure 6 we plot $H_d(G)$.

Figure 4. $dct(G)$ curves for floorplans with $n = 30$ or $n = 28$ cells. Each curve corresponds to a fixed (M, N) pair. The horizontal axis corresponds to the edge insertion probability p_0.

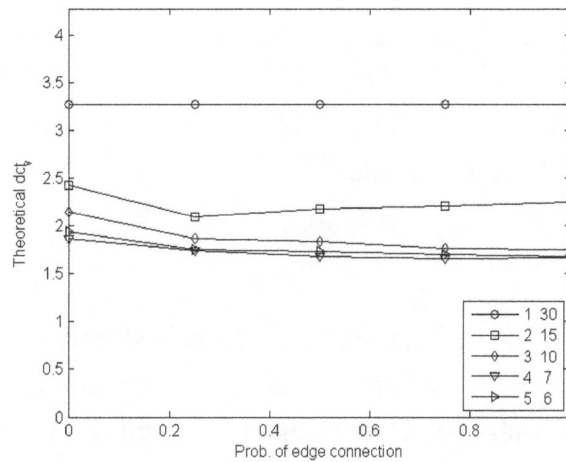

Figure 5. $dct_i(G)$ curves for floorplans with $n = 30$ or $n = 28$ cells. Each curve corresponds to a fixed (M, N) pair. The horizontal axis corresponds to the edge insertion probability p_0.

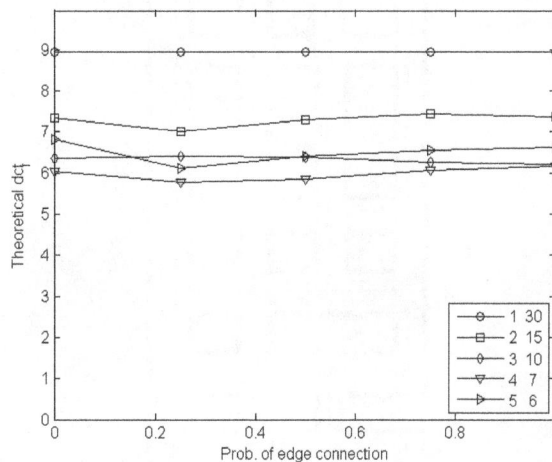

Figure 6. $H_d(G)$ curves for floorplans with $n = 30$ or $n = 28$ cells. Each curve corresponds to a fixed (M, N) pair. The horizontal axis corresponds to the edge insertion probability p_0.

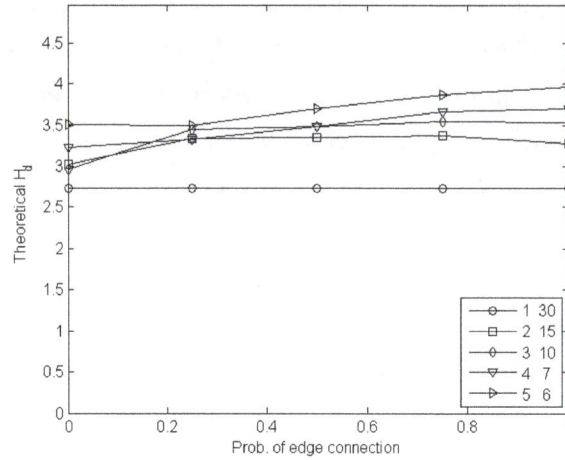

We can see in Figures 4 and 5 that both $dct(G)$ and $dct_i(G)$ are usually decreasing functions of the M/N ratio. However the cost of visibility $H_d(G)$ *increases* with M/N. This is due to the fact that, when the M/N ratio is low, G is closer to a path and there is less difference in the search schedules and capture times between dv-CR and di-CR. On the other hand, for high M/N ratio, G is closer to a grid, with a significantly increased ratio of edges to nodes (as compared to the low M/N, path-like instances). This, combined with the loss of information (visibility), results in $H_d(G)$ being an increasing function of M/N. The increase of $H_d(G)$ with p_0 can be explained in the same way, since increasing p_0 implies more edges and this makes the cops' task harder.

6.2. Experiments with Edge Games

Next we deal with $\overline{H}_d(G) = \frac{\overline{dct_i}(G)}{\overline{dct}(G)}$. We use graphs G obtained from *mazes* such as the one illustrated in Figure 7. Every *corridor* of the maze corresponds to an edge; corridor intersections correspond to nodes. The resulting graph G is also depicted in Figure 7. From G we obtain the line graph $L(G)$, to which we apply CADR to compute $dct(L(G)) = \overline{dct}(G)$ and PCS to compute $dct_i(L(G)) = \overline{dct_i}(G)$.

Figure 7. A maze and the corresponding graph.

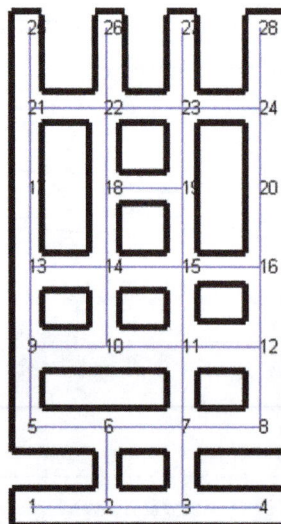

We use graphs of the same type as the ones of Section 6.1 but we now focus on the edge-to-edge movements of cops and robber. Hence from every G (obtained by a specific (M, N, p_0) triple) we produce the line graph $L(G)$, for which we compute $H_d(L(G))$ using the CADR and PCS algorithms. Once again we generate 50 graphs and present average $dct(G)$, $dct_i(G)$ and $H_d(G)$ results in Figures 8–10. These figures are rather similar to Figures 4–6, except that the increase of $\overline{H}_d(G)$ as a function of M/N is greater than that of $H_d(G)$. This is due to the fact that $L(G)$ has more nodes and edges than G, hence the loss of visibility makes the edge game significantly harder than the node game. There is one exception to the above remarks, namely the case $(M, N) = (1, 30)$; in this case both G and $L(G)$ are paths and $H_d(G)$ is essentially equal to $\overline{H}_d(G)$ (as can be seen by comparing Figures 6 and 10).

Figure 8. $\overline{dct}(G)$ curves for floorplans with $n = 30$ or $n = 28$ cells. Each curve corresponds to a fixed (M, N) pair. The horizontal axis corresponds to the edge insertion probability p_0.

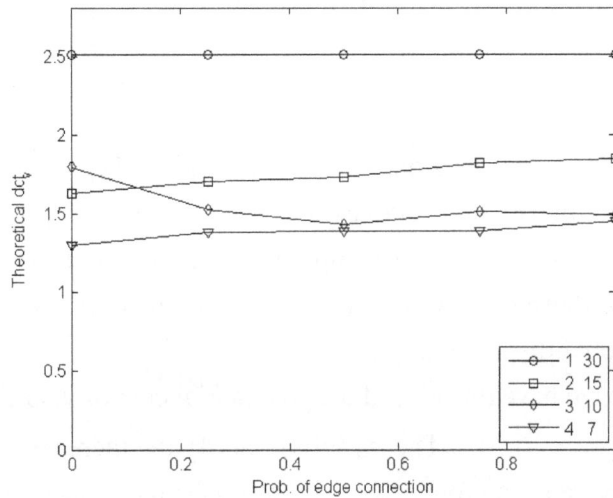

Figure 9. $\overline{dct}_i(G)$ curves for floorplans with $n = 30$ or $n = 28$ cells. Each curve corresponds to a fixed (M, N) pair. The horizontal axis corresponds to the edge insertion probability p_0.

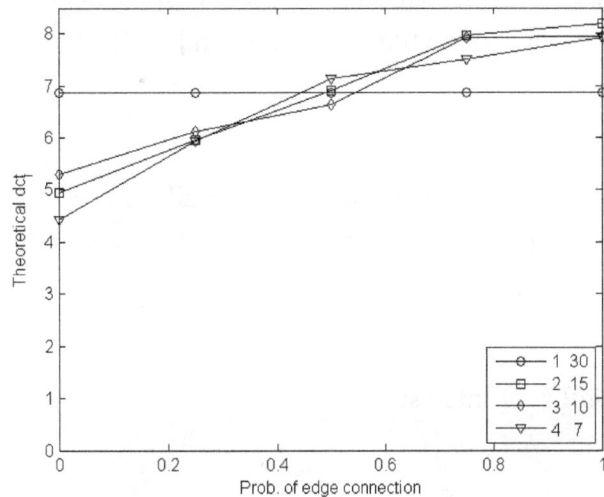

Figure 10. $\overline{H}_d(G)$ curves for floorplans with $n = 30$ or $n = 28$ cells. Each curve corresponds to a fixed (M, N) pair. The horizontal axis corresponds to the edge insertion probability p_0.

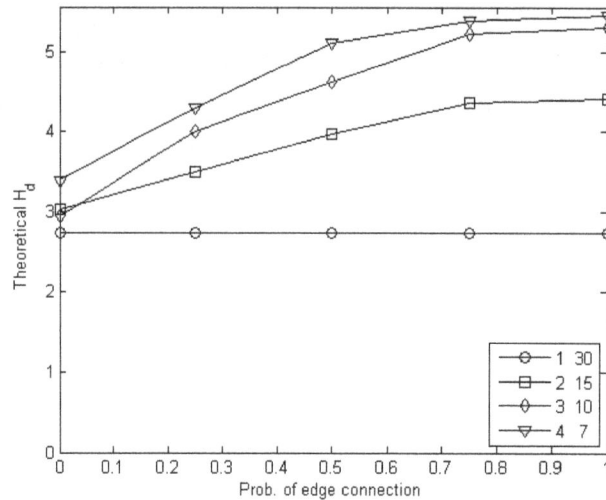

7. Conclusions

In this paper we have studied two versions of the cops and robber game: the one is played on the nodes of a graph and the other played on the edges. For each version, we studied four variants, obtained by changing the visibility and adversariality assumptions regarding the robber; hence we have a total of eight CR games. For each of these we have defined *rigorously* the corresponding optimal capture time, using game theoretic and probabilistic tools.

Then, for the node games we have introduced the adversarial cost of visibility $H(G) = \frac{ct_i(G)}{ct(G)}$ and the drunk cost of visibility $H_d(G) = \frac{dct_i(G)}{dct(G)}$. These ratios quantify the increase in difficulty of the CR game when the cop is no longer aware of the robber's position (this situation occurs often in mobile robotics).

We have defined analogous quantities ($\overline{H}(G) = \frac{\overline{ct_i}(G)}{\overline{ct}(G)}$, $\overline{H}_d(G) = \frac{\overline{dct_i}(G)}{\overline{dct}(G)}$) for the edge CR games.

We have studied analytically $H(G)$ and $H_d(G)$ and have established that both can get arbitrarily large. We have established similar results for $\overline{H}(G)$ and $\overline{H}_d(G)$. In addition, we have studied $H_d(G)$ and $\overline{H}_d(G)$ by numerical experiments which support both the game theoretic results of the current paper and the analytical computations of capture times presented in [9,37].

Author Contributions

Each of the three authors of the paper has contributed to all aspects of the theoretical analysis. The numerical experiments were designed and implemented by Athanasios Kehagias.

Conflicts of Interest

The authors declare no conflict of interest.

References

1. Chung, T.H.; Hollinger, G.A.; Isler, V. Search and pursuit-evasion in mobile robotics. *Auton. Robots* **2011**, *31*, 299–316.

2. Isler, V.; Karnad, N. The role of information in the cop-robber game. *Theor. Comput. Sci.* **2008**, *399*, 179–190.

3. Alspach, B. Searching and sweeping graphs: A brief survey. *Le Matematiche* **2006**, *59*, 5–37.

4. Bonato, A.; Nowakowski, R. *The Game of Cops and Robbers on Graphs*; AMS: Providence, RI, USA, 2011.

5. Fomin, F.V.; Thilikos, D.M. An annotated bibliography on guaranteed graph searching. *Theor. Comput. Sci.* **2008**, *399*, 236–245.

6. Nowakowski, R.; Winkler, P. Vertex-to-vertex pursuit in a graph. *Discret. Math.* **1983**, *43*, 235–239.

7. Dereniowski, D.; Dyer, D.; Tifenbach, R.M.; Yang, B. Zero-visibility cops and robber game on a graph. In *Frontiers in Algorithmics and Algorithmic Aspects in Information and Management*; Springer: Berlin, Germany, 2013; pp. 175–186.

8. Isler, V.; Kannan, S.; Khanna, S. Randomized pursuit-evasion with local visibility. *SIAM J. Discret. Math.* **2007**, *20*, 26–41.

9. Kehagias, A.; Mitsche, D.; Prałat, P. Cops and invisible robbers: The cost of drunkenness. *Theor. Comput. Sci.* **2013**, *481*, 100–120.

10. Adler, M.; Racke, H.; Sivadasan, N.; Sohler, C.; Vocking, B. Randomized pursuit-evasion in graphs. *Lect. Notes Comput. Sci.* **2002**, *2380*, 901–912.

11. Vieira, M.; Govindan, R.; Sukhatme, G.S. Scalable and practical pursuit-evasion. In Proceedings of the 2009 IEEE Second International Conference on Robot Communication and Coordination (ROBOCOMM'09), Odense, Denmark, 31 March–2 April 2009; pp. 1–6.

12. Gerkey, B.; Thrun, S.; Gordon, G. Parallel stochastic hill-climbing with small teams. In *Multi-Robot Systems. From Swarms to Intelligent Automata*; Springer: Dordrecht, Netherlands, 2005; Volume III, pp. 65–77.

13. Hollinger, G.; Singh, S.; Djugash, J.; Kehagias, A. Efficient multi-robot search for a moving target. *Int. J. Robot. Res.* **2009**, *28*, 201–219.

14. Hollinger, G.; Singh, S.; Kehagias, A. Improving the efficiency of clearing with multi-agent teams. *Int. J. Robot. Res.* **2010**, *29*, 1088–1105.

15. Lau, H.; Huang, S.; Dissanayake, G. Probabilistic search for a moving target in an indoor environment. In Proceedings of the 2006 IEEE/RSJ International Conference on Intelligent Robots and Systems, Beijing, China, 9–15 October 2006; pp. 3393–3398.

16. Sarmiento, A.; Murrieta, R.; Hutchinson, S.A. An efficient strategy for rapidly finding an object in a polygonal world. In Proceedings of the 2003 IEEE/RSJ International Conference on Intelligent Robots and Systems(IROS 2003), Las Vegas, NV, USA, 27–31 October 2003; Volume 2, pp. 1153–1158.

17. Hsu, D.; Lee, W.S.; Rong, N. A point-based POMDP planner for target tracking. In Proceedings of the 2008 IEEE International Conference on Robotics and Automation (ICRA 2008), Pasadena, CA, USA, 19–23 May 2008; pp. 2644–2650.

18. Kurniawati, H.; Hsu, D.; Lee, W.S. Sarsop: Efficient point-based POMDP planning by approximating optimally reachable belief spaces. In Proceedings of Robotics: Science and Systems, Zurich, Switzerland, 25–28 June 2008.

19. Pineau, J.; Gordon, G. POMDP planning for robust robot control. *Robot. Res.* **2007**, *28*, 69–82.

20. Smith, T.; Simmons, R. Heuristic search value iteration for POMDPs. In Proceedings of the 20th Conference on Uncertainty in Artificial Intelligence, Banff, Canada, 7–11 July 2004; pp. 520–527.

21. Spaan, M.T.J.; Vlassis, N. Perseus: Randomized point-based value iteration for POMDPs. *J. Artif. Intel. Res.* **2005**, *24*, 195–220.

22. Hauskrecht, M. Value-function approximations for partially observable Markov decision processes. *J. Artif. Intel. Res.* **2000**, *13*, 33–94.

23. Littman, M.L.; Cassandra, A.R.; Kaelbling, L.P. *Efficient Dynamic-Programming Updates in Partially Observable Markov Decision Processes*; Technical Report CS-95-19; Brown University, Providence, RI, USA, 1996.

24. Monahan, G.E. A survey of partially observable Markov decision processes: Theory, models, and algorithms. *Manag. Sci.* **1982**, *28*, 1–16.

25. Canepa, D.; Potop-Butucaru, M.G. Stabilizing Flocking Via Leader Election in Robot Networks. In Proceedings of the 9th International Symposium on Stabilization, Safety, and Security of Distributed Systems (SSS 2007), Paris, France, 14–16 November 2007; pp. 52–66.

26. Gervasi, V.; Prencipe, G. Robotic Cops: The Intruder Problem. In Proceedings of the 2003 IEEE Conference on Systems, Man and Cybernetics (SMC 2003), Washington, DC, USA, 5–8 October 2003; pp. 2284–2289.

27. Prencipe, G. The effect of synchronicity on the behavior of autonomous mobile robots. *Theory Comput. Syst.* **2005**, *38*, 539–558.

28. Dudek, A.; Gordinowicz, P.; Pralat, P. Cops and robbers playing on edges. *J. Comb.* **2013**, *5*, 131–153.

29. Kuhn, H.W. Extensive games. *Proc. Natl. Acad. Sci. USA* **1950**, *36*, 570–576.

30. Bonato, A.Y.; Macgillivray, G. *A General Framework for Discrete-Time Pursuit Games*, preprint.

31. Hahn, G.; MacGillivray, G. A note on k-cop, l-robber games on graphs. *Discret. Math.* **2006**, *306*, 2492–2497.

32. Berwanger, D. *Graph Games with Perfect Information*, preprint.

33. Mazala, R. Infinite games. In *Automata, Logics and Infinite Games*; Springer-Verlag: Berlin, German, **2002**, *2500*, 23–38.

34. Aigner, M.; Fromme, M. A game of cops and robbers. *Discret. App. Math.* **1984**, *8*, 1–12.

35. Osborne, M.J. *A Course in Game Theory*; MIT Press: Cambridge, MA, USA, 1994.

36. Puterman, M.L. *Markov Decision Processes: Discrete Stochastic Dynamic Programming*; John Wiley & Sons, Inc.: New York, NY, USA, 1994.

37. Kehagias, A.; Prałat, P. Some remarks on cops and drunk robbers. *Theor. Comput. Sci.* **2012**, *463*, 133–147.

38. De la Barrière, R.P. *Optimal Control Theory: A Course in Automatic Control Theory*; Dover Pubns: New York, NY, USA, 1980.

39. Eaton, J.H.; Zadeh, L.A. Optimal pursuit strategies in discrete-state probabilistic systems. *Trans. ASME Ser. D J. Basic Eng.* **1962**, *84*, 23–29.

40. Howard, R.A. *Dynamic Probabilistic Systems, Volume Ii: Semi-Markov and Decision Processes*;
     Dover Publications: New York, NY, USA, 1971.

41. Raghavan, T.E.S.; Filar, J.A. Algorithms for stochastic games—A survey. *Math. Methods Oper.*
     Res. **1991**, *35*, 437–472.

9

Design and Simulation of Two Robotic Systems for Automatic Artichoke Harvesting

Domenico Longo [1,]* and Giovanni Muscato [2]

[1] Department of Agri-Food and Environmental Systems Management (DiGeSA), Section of Mechanics and Mechanization, University of Catania, Via S. Sofia 100, Catania 95123, Italy

[2] Electric, Electronics and Computer Engineering Department (DIEEI), University of Catania, Viale A. Doria 6, Catania 95125, Italy; E-Mail: gmuscato@dieei.unict.it

* Author to whom correspondence should be addressed; E-Mail: dlongo@unict.it

Abstract: The target of this research project was a feasibility study for the development of a robot for automatic or semi-automatic artichoke harvesting. During this project, different solutions for the mechanical parts of the machine, its control system and the harvesting tools were investigated. Moreover, in cooperation with the department DISPA of University of Catania, different field structures with different kinds of artichoke cultivars were studied and tested. The results of this research could improve artichoke production for preserves industries. As a first step, an investigation on existing machines has been done. From this research, it has been shown that very few machines exist for this purpose. Based also on previous experiences, some proposals for different robotic systems have been done, while the mobile platform itself was developed within another research project. At the current stage, several different configurations of machines and harvesting end-effectors have been designed and simulated using a 3D CAD environment interfaced with Matlab®. Moreover, as support for one of the proposed machines, an artificial vision algorithm has been developed in order to locate the artichokes on the plant, with respect to the robot, using images taken with a standard webcam.

Keywords: artichoke harvesting; agricultural robotics; artificial vision; outdoor autonomous robot

1. Introduction

In recent years agriculture has received the focus of different research groups that operate in the robotics and automation field. Moreover, electronics, informatics and automation solutions have been proposed as commercial products by most of agriculture machine and tools suppliers. Nevertheless, in-field fully autonomous operations yet have some barriers to overcome in order to become relatively common and cheap as well as reliable, in comparison to the great number of applications in the manufacturing industry.

The main reasons for this delay is due to the fact that agricultural tasks are carried out in an outdoor environment, often really unstructured, dynamic and most of the time heavily conditioned by water, mud, wind, light, dust, chemicals and so on. Moreover, the actual cost of a robotic system is usually too high with respect to the other production costs and with respect to the standard season operators' salary. Despite these premises, different solutions exist for some specific agricultural operations. In [1,2] a robotic system for vineyard and greenhouses operation, like transportation and spraying, is reported. Several other examples of robotic harvesting systems have been developed and reported in [3–5]. However not so many commercial systems exist and most of these are simply research laboratory prototypes. In [6], an interesting review on the fruit harvesting problem and the possible automatic solutions is reported. Among the harvesting robots that have been developed, it is possible to cite the Orange Picking Robots [7] developed by the University of Catania, where a detailed study on the vision processing analysis was carried out and several testing of the picking mechanism were developed. In [8,9], two in-depth reviews about artificial vision applications to the agricultural problem and useful algorithm are reported, while in [10] a possible solution to the fruit detection problem has been implemented and tested. Another interesting research activity has been carried out by Wageningen University that performed several studies and field tests for a cucumber harvesting robot [11]. As regard harvesting, many applications concentrated in greenhouse environment, where it is simpler to move the harvesting system and to locate vegetables, as in the case of strawberry harvesting [12]. Only a few works concentrate on vegetable harvesting in outdoor fields, as in the case of radicchio harvesting [13]. Other interesting applications of robotics in agriculture include precision spraying and precision fertilization [14], inspection and treatment of plants [15], selective herbicide application [16], just to name a few.

As far as we know, there is only one example of a machine for artichoke harvesting. The aim of the BIOCARD EU FP6 STREP project [17,18] was to demonstrate the economical and technical feasibility of a global process to improve annual lifecycle Cynara cardunculus exploitation for energy applications. Among the different activities of that project, the development of specific machinery able to harvest the capitula and separate directly in the field seeds from cynara biomass was addressed. Concerning automatic artichoke harvesting for fresh vegetable market, our research work represents a preliminary step toward the introduction of robotics into this specific field. In particular the main peculiarities of our approach are from one side the design and development of an automated robotic system for harvesting and on the other side the fact that the research was carried out working concurrently with the growers and with the experts in horticulture, to select suitable plants that could make the process of harvesting easier.

2. Investigation about Artichokes Field Design and Current Harvest Methodology

The purpose of this research project was a feasibility study for the development of an automatic or semi-automatic machine for artichokes harvesting. This kind of machine will work on specific cultivations for industrial use of the artichokes, like for preserves production. These kinds of cultivations are more indicated for maximum productivity with minimum interest to artichoke appearances and could be selected in a suitable way, in order to be compatible for the automatic harvesting while maintaining organoleptics properties. This task is easier with respect to many other crops because most artichoke cultivars have an annual or bi-annual life cycle. Moreover, suitable cultivars that have specific peculiarities for industrial use and that could simplify automatic harvesting have been tested.

In our study case, with the help of the department DISPA of the University of Catania, two cultivars have been selected for their properties: Harmony F1 hybrid and Madrigal F1 hybrid. Moreover, the field structure and density have been designed in a suitable way, in order to allow a small vehicle to move along the rows. The first field structure was realized with a single row of plants, with different densities between 0.4 m and 0.7 m. The row distance has been kept at 1.4 m for each density. The second field structure, shown in Figure 1, is realized with doubled interlaced rows, with different densities between 0.55 m and 1 m. The distance between the rows is kept at 1.4 m.

Figure 1. Some field structure used during test: doubled interlaced rows.

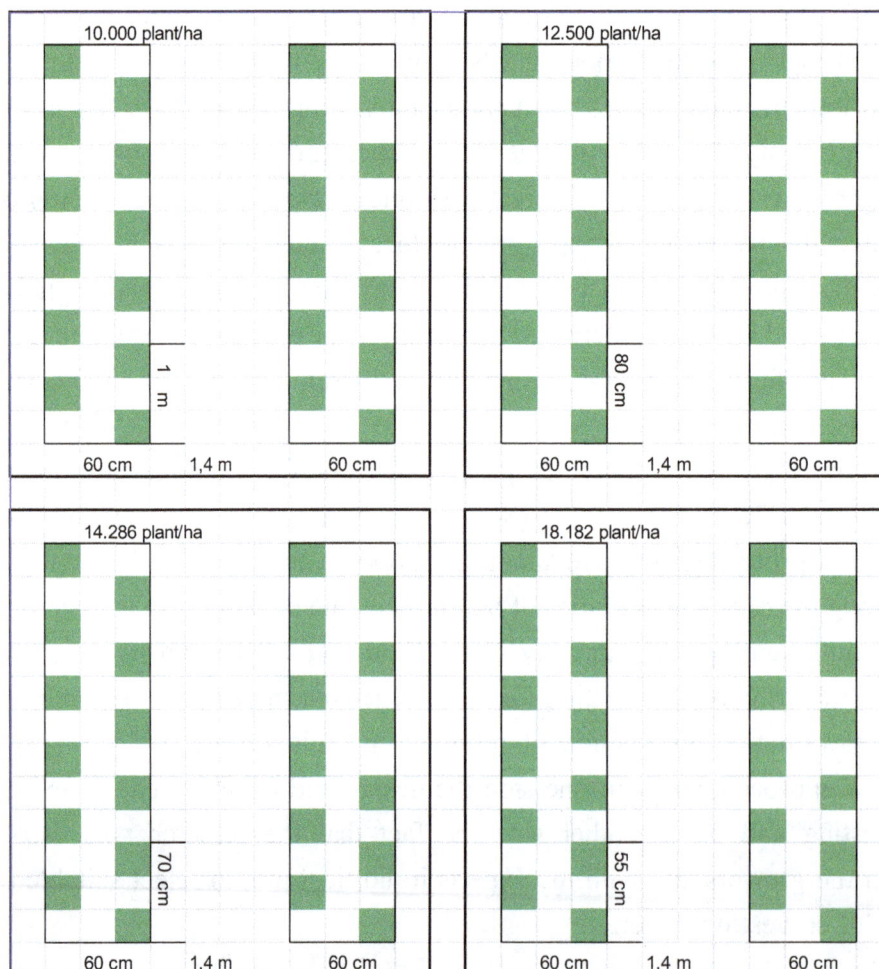

Until now, artichokes are manually harvested and collected inside wooden or plastic-made boxes for retail business. The only difference with harvesting for industrial business is that artichokes are collected directly inside trucks or small tractors, like that in Figure 2. Manual harvest is often made by using some support machines, mainly for all the collected artichokes. From some statistical analysis in the Sicily area, the harvest cost is about 20% of the total production cost. In order for automatic harvesting to be convenient, a suitable machine with good harvesting rate must be developed. A typical harvesting rate of 10 workers is about 30,000 artichokes/day, with a typical 8 h long working time, which is 6.25 artichokes each minute per worker, so any kind of machine must at least have a similar harvesting rate.

Figure 2. A small tractor used for manual artichokes harvesting.

3. Machines Design and Simulations

On the basis of several field observations and measurements and following discussions with local experts, two different harvesting machines have been designed and simulated. The two machines are based on different harvesting methodologies and each one has its own advantages and drawbacks. Each of the two uses a mobile platform that relies on two rubber tracks. It is possible to use electrical motors or thermal engine ones. A suitable small electrical tracked vehicle was developed and tested by the authors along another research project and is reported in [1,2]. This is narrow enough to be able to move between the plant rows. The two machines have been simulated in a 3D CAD environment, that can be interfaced to Matlab® packages. In order to implement a more realistic simulation, first of all a graphical representation of an artichoke with trunks, leaves and immature flowers (Figure 3) was prepared. Each artichoke head has a known geometrical position in the simulated field (Figure 4) and this property has been used during the simulations in order to cut or pick-up the artichoke itself with the simulated machine.

Figure 3. The simulated artichoke plant.

Figure 4. The simulated artichokes field.

The first proposed machine uses a dual manipulator configuration. One manipulator has three degrees of freedom (DOF), while the other has only two DOF. Once the machine has reached a suitable position to use the cutting tool on the first manipulator, the system can adjust the position of the second manipulator in order to pick another artichoke without moving the mobile base. Each manipulator has an end-effector capable to cut the artichoke and to drop it directly inside a specific box or on a conveyor belt (Figure 5). Each DOF could be pneumatically or hydraulically actuated. The main peculiarity of this machine is that it can harvest artichokes with high precision with the help of an artificial vision system described in next paragraph. However, due to the mechanical delays, the system can not be fast enough to satisfy the required minimum harvesting rate. The plant will not be damaged too much by the harvesting procedure, but, due to some possible faults in the artificial vision system, some artichokes will not be seen and consequently not harvested.

Figure 5. The robot for artichokes selective harvesting: (**a**) and (**b**) two different views of the first proposal and (**c**) a detailed view of the gripping tool.

(a)

(b)

(c)

The control architecture should be composed by four main blocks: the vision system, the mobile platform position control system, the manipulators control system and the main supervising and coordinating control system. Once the supervising and coordinating block is signaled from the vision system that an artichoke is found in the image in a suitable workspace, it stops the mobile base at the given position. Then sends the artichoke coordinates, relative to the base, given by the vision system to the first manipulator. At the same time, if another artichoke is found on the operative space of the second manipulator, also this is commanded to move and harvest. Then the procedure is repeated until another artichoke is found in the image or the end of the corridor is reached.

The second type of machine uses a very large cutting floor with moving blades (Figure 6). This kind of machine will cut the plant at a suitable height. The height value will be estimated time by time before starting the harvesting session and regulated accordingly. This machine could be fast enough, but it could damage plants and, if the height of the cutting floor is not well regulated, the artichokes could be also damaged. In particular also some foliage will be cut and collected during the harvesting of the artichokes. Most of the separation can be performed by using some selecting tools onboard the machine. The remaining foliage can be easily removed during the subsequent factory processing phases. The damage caused on the plants is not relevant because the life cycle of these cultivars is annual or bi-annual. For all the proposed machines, the 3D environments have been interfaced with Matlab® for simulating their control system.

Figure 6. (a) The robot for artichokes mass-harvesting. **(b)** Machine details of the cutting floor.

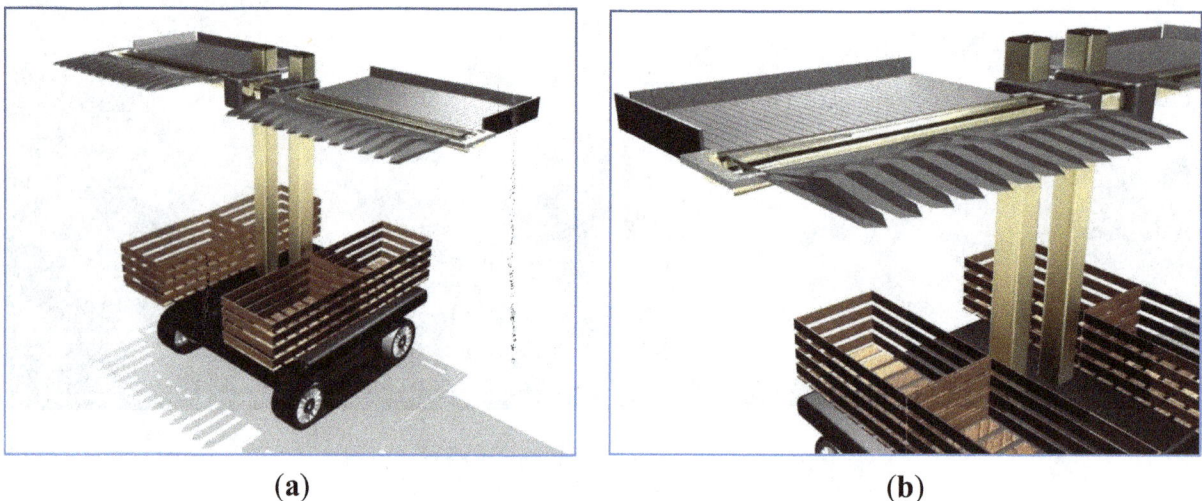

(a)   (b)

4. Artificial Vision Algorithm

The first proposed machine uses two standard cartesian manipulators, one with two and the second with three linear joints to pick artichokes with a suitable end-effector (cutter). In order to perform a correct positioning of the cutter with respect to the artichokes, an artificial vision algorithm based on images acquired with a standard webcam mounted on-board the robot, was adopted. All images are taken as top view of the field at a mean distance of about 0.6 m from the top of the plant. For development purposes, the algorithm has been implemented in a Matlab® environment using the Image Processing Toolbox; it has been tested with static images from the camera. In Figure 7, a typical image taken with the camera is shown. This image is then converted in black and white (Figure 8) and scaled down in resolution, in order to implement a very fast recognition algorithm.

Figure 7. Camera picture of an on-field artichoke.

Figure 8. Black and white representation.

4.1. Details of the Image Processing Algorithm

Different methodologies have been used to obtain better results. Among the others, erosion-diffusion techniques, correlation with sample images, Artificial Neural Networks, co-occurrence matrix, cluster symmetry, wavelets and so on. Due to the fact that artichokes have quite the same color of the image background (mainly leafs but also terrain with weeds), image colors are not very useful to identify them. On the other side, searching for high-symmetry clusters in the image could help. In case the picture contains portion of terrain, these areas are considered as noise. Using a combination of the previous methodologies as a sequence of image filters, it is possible to obtain a high true positive rate in each image. Best results have been obtained with a preliminary black and white conversion, followed by an erosion-diffusion step. After that, the algorithm searches for high-symmetry regions. To compute such regions, the Specular Image method was applied. In the main image, a moving sub-image denoted as "kernel" is established around a specific pixel on the image and then the left and right parts of this kernel are compared each other as well as the upper and lower part. In steps of 10 pixel, a new kernel is defined around the next 10th pixel and so on for all the main image pixels. In order to maintain a suitable speed for the algorithm the image of was scaled down in resolution to 512×384 pixels. Since for all the images taken in the used experimental conditions, artichoke commonly cover an area of about 60×60 pixels, a kernel dimension of 105×105 pixels was chosen as a good compromise. When the two parts of the kernel have about 3000 similar pixels the image is considered has having a high symmetry.

In all those regions, patterns are then classified using GLCM (Gray-Level Co-occurrence Matrix) [19] and Wavelets methodology [20] in order to filter out most of the false positive targets. These two analyses aim to obtain some kind of digital signature of the artichoke in its different shape, light conditions and so on. From different trials, is has been shown that leafs and other kind of noises on the images have different signatures in the sense of GLCM and Wavelets analysis. Each of the two methods are first performed on different samples images containing typical artichoke, leafs and terrain in real conditions. These samples images have typical dimensions of 51×51 pixels. After the two methods on sample image with suitable pre-filters have been applied in order to find out thresholds for all the different parameters, auto and cross correlation operations are performed between samples images and related parameters have been computed and compared.

As regards the GLCM analysis, the *Contrast* output parameter is considered, as it was found as the only parameter able to discriminate an artichoke from leaves and terrain. Then, the Entropy and Standard Deviation of the same sub-image are computed. From a series of experimental trials on real images, it was found that the sub-image can be considered a true positive, when the parameters move in the ranges described in Table 1.

Table 1. Ranges for the used parameters of the Gray-Level Co-occurrence Matrix (GLCM) algorithm.

Parameter	Minimum value	Maximum value
Contrast	0.0755	0.16
Entropy	0.2745	0.33
Standard Deviation	0.2045	0.24

Regarding the second analysis, a two dimensional wavelet function decomposition is then used. The best results have been obtained using a wavelet function "db8" of the Daubechies family with three levels of decomposition. After this operation, always applied over sample images, the *approximation* parameter have been rejected while all the three *detail* parameters have been collected in the "signature" matrix. A specific kernel image with "high symmetry" is considered a true target when the correlation between the two signature matrixes is greater than 0.87.

4.2. Results of the Image Processing Algorithm

In Figure 9 the original image with some information overlapped about the real artichoke position with respect to the robot, is shown. Green squares are related to true positive artichokes coordinate points, while red squares indicate false positive. The robot will then reach only the true positive points. During different tests, the vision algorithm has shown good capabilities to detect true and false positive

(85% to detect true positive and 76% to detect false positive) on a set of about 20 typical artichoke images taken in a real field in sunny conditions.

Figure 9. Localization data computed by the artificial vision algorithm, on the original image.

5. Results

In order to get an estimation of the performance of the two solutions, *ad-hoc* simulations were done because each of the two machines uses a different approach.

As regards the mass-harvesting machine, the main problem is the different height of artichokes on the plant, even if all of them are at the same maturity level. The real case is even worst, because of the presence of artichokes not ready to be harvested (too small or not mature). For this kind of simulation, several measurements were done on real fields to get a statistical distribution of the positions and heights of the artichokes. In Figure 10 the mean of the heights of different order artichokes (main stem/bud, first order and second order stems/buds are considered [21]) that are at the same time on the plants, for nine different repetitions, is shown. Each repetition was made in different days, normally with one week interval. In this experiment, the cultivar Madrigal F1 in simple rows was considered. These measurements were also used to implement the Matlab® simulations for the control system of the two machines, while operating in the 3D CAD environment. In these simulations the position distribution of the artichokes and the recognition rate of the vision algorithm (for the selective harvesting robot) were also taken into account. The planar position was reproduced from the measurements on a real field, while the height of the buds was modeled using a Gaussian distribution with the mean and variance get from the field. In this way it was possible to estimate the harvesting time for each robot, as reported in Table 2.

Figure 10. Mean height in centimeters for artichokes of the cultivar Madrigal F1 in simple rows for nine different manual harvesting sessions.

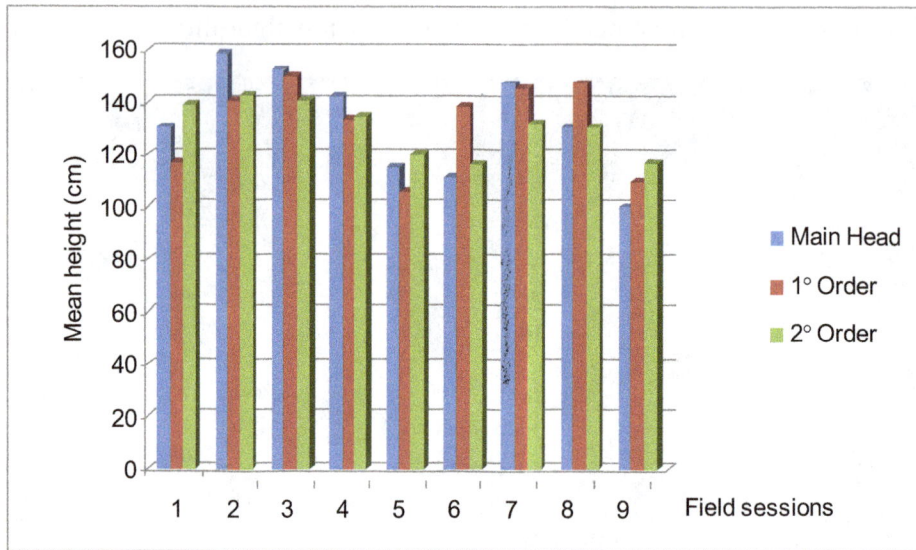

Table 2. Harvesting performance of the two proposed solutions derived from the Matlab® simulations.

	Selective Harvesting robot	Mass Harvesting robot
Speed	2 m/s (Arm)–0.2 m/s (Vehicle)	0.2 m/s (Vehicle)
Productivity	0.13 Artichokes/s	1.2 Artichokes/s
Productivity per hour	480 Artichokes/h	4,320 Artichokes/h
TOTAL Productivity (Working day)	3,840 Artichokes/8 h	34,560 Artichokes/8 h
Weight harvested (Unit weight of typical artichoke 200 g)	96 kg/h	864 kg/h

In order to obtain an estimation of the efficiency of the two machines, in terms of achieved useful yield, a simulation of the proposed harvesting methodologies has been also done in the fields by manually harvesting artichokes using the two proposed methodologies. Operators were trained about the two protocols. With the mass-harvesting protocol, operators harvested all the artichokes that are above a certain height, established before the session. The harvesting days were chosen by skilled operators in order to maximize the presence on the whole field of artichokes to be harvested. Artichokes that are at the cutting level or above the cutting level, but not ready to be harvested, were tagged as "damaged". The same experimental flow applies to the selective-harvesting protocol.

After the harvesting phase, collected artichokes were counted and cumulative data for different harvesting sessions are shown in the graphs in Figures 11 and 12.

It can be observed that the mass harvesting robot has a good productivity, comparable to the given requirements. However, the percentage of damaged artichokes is too high to be considered suitable for the intended purposes. The selective harvesting has a small rate of damaged artichokes, but is too much slow, complex and expensive to be really implemented in the field.

Figure 11. Total harvested and damaged artichokes for the selective harvesting robot.

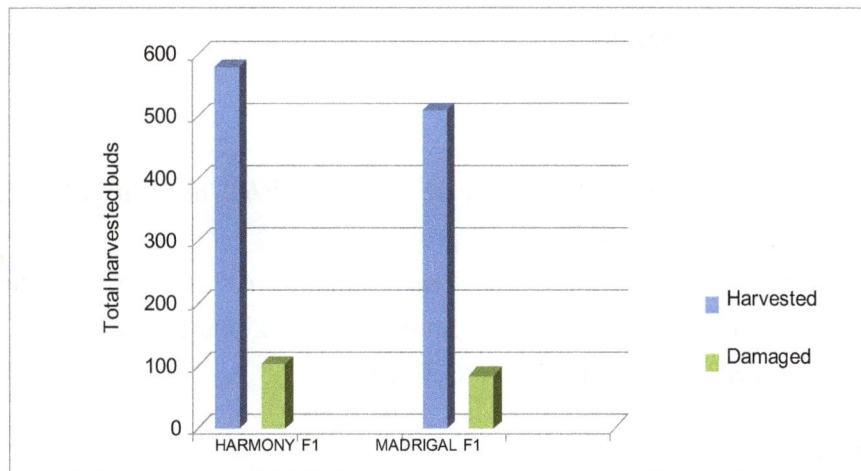

Figure 12. Total harvested and damaged artichokes for the mass harvesting robot.

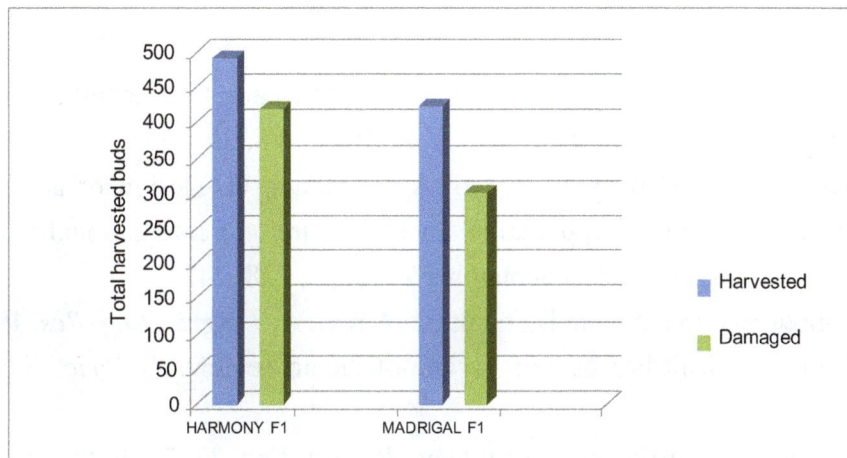

6. Conclusions

The problem of artichoke harvesting has been investigated by means of literature search, in-field tests and manually simulating the robotic harvesting. Exploiting the fact that artichoke plants are mainly annual or bi-annual, it is possible to easily design the field in order to be compatible with mechanical harvesting operations, by means of a small autonomous vehicle with suitable manipulators or tools. Also artichoke cultivars have been selected to maximize mass production and to have plant field more compatible with mechanical harvesting.

With the help of a 3D CAD environment, two different approaches to the artichoke harvesting problem have been proposed. By using this methodology, some advantages and drawbacks of the proposed solutions have been fully analyzed, taking into account field observation on real plants and local expertise. Moreover, as a support for one of the proposed solution, an artificial vision algorithm has been developed and tested in order to recognize artichokes on the plant to allow the robot to pick them. From the obtained results it follows that further work is needed before building and testing a real prototype, since the obtained performance for those systems is still not satisfactorily. In any case the proposed approaches give us a suitable tool and a strong basis for the development, simulation and test of new solutions. This work represents a preliminary study about possible solutions for automatic artichoke harvesting. In the next phases of the research, further collecting criteria and performance requirements will be certainly assessed.

Acknowledgements

The authors would like to thank G. Mauromicale of the DISPA University of Catania and his staff for the helpful cooperation established during the project.

This project has been funded by the Regione siciliana, Assessorato Agricoltura e Foreste, Dipartimento Interventi Infrastrutturali, project: "Studio progettazione e costruzione di un robot per la raccolta automatizzata dei capolini di carciofo—RACAR".

Conflicts of Interest

The authors declare no conflict of interest.

References

1. Longo, D.; Pennisi, A.; Bonsignore, R.; Schillaci, G.; Muscato, G. A small autonomous electrical vehicle as partner for heroic viticulture. *Acta Hort.* **2013**, *978*, 391–398.
2. Longo, D.; Muscato, G.; Caruso, L.; Conti, A.; Schillaci, G. Design of a Remotely Operable Sprayer for Precision Farming Application. In Proceedings of International Conference Ragusa SHWA2012, Ragusa Ibla, Italy, 3–6 September 2012; pp. 238–242.
3. Tillet, N.D. Robotic manipulators in horticulture: A review. *J. Agric. Eng. Res.* **1993**, *55*, 89–105.
4. Sarig, Y. Robotics for fruit harvesting: A state-of-the-art review. *J. Agric. Eng. Res.* **1993**, *54*, 265–280.
5. Sanders, K.F. Orange harvesting systems review. *Biosyst. Eng.* **2005**, *90*, 115–125.
6. Li, P.; Lee, S.; Hsu, H.-Y. Review on fruit harvesting method for potential use of automatic fruit harvesting systems. *Proc. Eng.* **2011**, *23*, 351–366.
7. Muscato, G.; Prestifilippo, M.; Abbate, N.; Rizzuto, I. A prototype of an orange picking robot: Past history, the new robot and experimental results. *Ind. Robot. Int. J.* **2005**, *32*, 128–138.
8. Kapach, K.; Barnea, E.; Mairon, R.; Edan, Y.; Ben-Shahar, O. Computer vision for fruit harvesting robots—State of the art and challenges ahead. *Int. J. Comput. Vis. Robot.* **2012**, *3*, 4–34.
9. Jimenez, A.R.; Ceres, R.; Pons, J.L. A survey of computer vision methods for locating fruit on trees. *Trans. Am. Soc. Agric. Eng.* **2000**, *43*, 1911–1920.
10. Schillaci, G.; Pennisi, A.; Franco, F.; Longo, D. Detecting Tomato Crops in Greenhouses Using a Vision Based Method. In Proceedings of International Conference Ragusa SHWA2010, Ragusa Ibla, Italy, 3–6 September 2012; pp. 252–258.
11. Van Henten, E.J.; Hemming, J.; van Tuijl, B.A.J.; Kornet, J.G.; Meuleman, J.; Bontsema, J.; van Os, E.A. An autonomous robot for harvesting cucumbers in greenhouses. *Auton. Robots* **2002**, *13*, 241–258.

12. Hayashi, S.; Ota, T.; Kubota, K.; Ganno, K.; Kondo, N. Robotic Harvesting Technology for Fruit Vegetables in Protected Horticultural Production. In Proceedings of the Information and Technology for Sustainable Fruit and Vegetable Production, FRUITIC 05, Montpellier, France, 12–16 September 2005.

13. Foglia, M.M.; Reina, G. Agricultural robot for radicchio harvesting. *J. Field Robot.* **2006**, *23*, 363–377.

14. Belforte, G.; Deboli, R.; Gay, P.; Piccarolo, P.; Ricauda Aimonino, D. Robot design and testing for green house applications. *Biosyst. Eng.* **2006**, *95*, 309–321.

15. Acaccia, G.M.; Michelini, R.C.; Molfino, R.M.; Razzoli, R.P. Mobile Robots in Greenhouse Cultivation: Inspection and Treatment of Plants. In Proceedings of ASER 2003, 1st International Workshop on Advances in Service Robotics, Bardolino, Italy, 13–15 March 2003.

16. Lee, W.S.; Slaughter, D.C.; Giles, D.K. Robotic weed control system for tomatoes. *Precis. Agric.* **1999**, *1*, 95–113.

17. Pari, L.; Sissot, F.; Giannini, E. European Union Research Project Biocard: Mechanization Activities Results. In Proceedings of the Technology and Management to Increase the Efficiency in Sustainable Agricultural System Conference, Rosario, Argentina, 1–4 September 2009.

18. Pari, L.; Civitarese, V.; Assirelli, A.; del Giudice, A. Prototype for Cynara Cardunculus Capitula Threshing and Biomass Windrowing. In Proceedings of the 17th European Biomass Conference & Exibition—From Research to Industry and Markets, Hamburg, Germany, 29 June–3 July 2009; Volume 1, pp. 262–267.

19. Haralick, R.M.; Shanmugan, K.; Dinstein, I. Textural features for image classification. *IEEE Trans. Syst. Man Cybern.* **1973**, *3*, 610–621.

20. Mallat, S. A theory for multiresolution signal decomposition: The wavelet representation. *IEEE Pattern Anal. Mach. Intell.* **1989**, *11*, 674–693.

21. Archontoulis, S.V.; Struik, P.C.; Vos, J.; Danalatos, N.G. Phenological growth stages of Cynara cardunculus: Codification and description according to the BBCH scale. *Ann. Appl. Biol.* **2010**, *156*, 253–270.

Permissions

List of Contributors

Gareth Edwards
Department of Engineering, University of Aarhus, Blichers Allé 20, Tjele 8830, Denmark

Martin P. Christiansen
Department of Engineering, University of Aarhus, Blichers Allé 20, Tjele 8830, Denmark

Dionysis D. Bochtis
Department of Engineering, University of Aarhus, Blichers Allé 20, Tjele 8830, Denmark

Claus G. Sørensen
Department of Engineering, University of Aarhus, Blichers Allé 20, Tjele 8830, Denmark

Zachary Klaassen
Department of Surgery, Section of Urology, Medical College of Georgia, Georgia Regents University, Augusta, GA 30912, USA

Qiang Li
Department of Surgery, Section of Urology, Medical College of Georgia, Georgia Regents University, Augusta, GA 30912, USA

Rabii Madi
Department of Surgery, Section of Urology, Medical College of Georgia, Georgia Regents University, Augusta, GA 30912, USA

Martha K. Terris
Department of Surgery, Section of Urology, Medical College of Georgia, Georgia Regents University, Augusta, GA 30912, USA

Henrik Andreasson
Centre of Applied Autonomous Sensor Systems (AASS), Örebro University, 70182 Örebro, Sweden

Jari Saarinen
Centre of Applied Autonomous Sensor Systems (AASS), Örebro University, 70182 Örebro, Sweden

Marcello Cirillo
Centre of Applied Autonomous Sensor Systems (AASS), Örebro University, 70182 Örebro, Sweden

Todor Stoyanov
Centre of Applied Autonomous Sensor Systems (AASS), Örebro University, 70182 Örebro, Sweden

Achim J. Lilienthal
Centre of Applied Autonomous Sensor Systems (AASS), Örebro University, 70182 Örebro, Sweden

Kjeld Jensen
Faculty of Engineering, University of Southern Denmark, Campusvej 55, 5230 Odense M, Denmark

Morten Larsen
Conpleks Innovation, Fælledvej 17, 7600 Struer, Denmark

Søren H. Nielsen
Faculty of Engineering, University of Southern Denmark, Campusvej 55, 5230 Odense M, Denmark

Leon B. Larsen
Faculty of Engineering, University of Southern Denmark, Campusvej 55, 5230 Odense M, Denmark

Kent S. Olsen
Faculty of Engineering, University of Southern Denmark, Campusvej 55, 5230 Odense M, Denmark

Rasmus N. Jørgensen
Institute of Engineering, Aarhus University, Nordre Ringgade 1, 8000 Aarhus, Denmark

Niccoló Tosi
Department of Mechanical Engineering, KU Leuven, Celestijnenlaan 300, 3001 Heverlee, Belgium
CEA, LIST, Interactive Robotics Laboratory, PC 178, 91191 Gif sur Yvette Cedex, France

Olivier David
CEA, LIST, Interactive Robotics Laboratory, PC 178, 91191 Gif sur Yvette Cedex, France

Herman Bruyninckx
Department of Mechanical Engineering, KU Leuven, Celestijnenlaan 300, 3001 Heverlee, Belgium
Department of Mechanical Engineering, Eindhoven University of Technology, Eindhoven, The Netherlands

Christian Dondrup
School of Computer Science, University of Lincoln, Brayford Pool, LN6 7TS Lincoln, UK

Nicola Bellotto
School of Computer Science, University of Lincoln, Brayford Pool, LN6 7TS Lincoln, UK

Marc Hanheide
School of Computer Science, University of Lincoln, Brayford Pool, LN6 7TS Lincoln, UK

Kerstin Eder
Department of Computer Science, University of Bristol, Merchant Venturers Building, Woodland Road, Clifton, BS8 1UB Bristol, UK

Ute Leonards
School of Experimental Psychology, University of Bristol, 12A Priory Road, Clifton, BS8 1TU Bristol, UK

Peter Won
Mechanical and Mechatronics Engineering Department, University of Waterloo, Waterloo, ON, N2L 3G1, Canada

Mohammad Biglarbegian
School of Engineering, University of Guelph, Guelph, ON, N1G 2W1, Canada

William Melek
Mechanical and Mechatronics Engineering Department, University of Waterloo, Waterloo, ON, N2L 3G1, Canada

Athanasios Kehagias
Department of Electrical and Computer Engineering, Aristotle University, GR 54248, Thessaloniki, Greece

Dieter Mitsche
Laboratoire J. A. Dieudonné, UMR CNRS-UNS No 7351, Université de Nice Sophia-Antipolis, Pa0rc Valrose 06108 Nice Cedex 2, France

Paweł Prałat
Department of Mathematics, Ryerson University, 350 Victoria St., Toronto, ON, M5B 2K3, Canada

Domenico Longo
Department of Agri-Food and Environmental Systems Management (DiGeSA), Section of Mechanics and Mechanization, University of Catania, Via S. Sofia 100, Catania 95123, Italy

Giovanni Muscato
Electric, Electronics and Computer Engineering Department (DIEEI), University of Catania, Viale A. Doria 6, Catania 95125, Italy